DISTANT ENCOUNTERS

OTHER BOOKS BY MARK WASHBURN

NONFICTION
Mars at Last
In the Light of the Sun

NOVELS
The Armageddon Game
The Omega Threat
Nightwind

DISTANT ENCOUNTERS

THE EXPLORATION OF JUPITER AND SATURN

MARK WASHBURN

Illustrations by Susan Stillman

HARCOURT BRACE JOVANOVICH, PUBLISHERS
San Diego   New York   London

Copyright © 1983 by Mark Washburn

All rights reserved. No part of this publication
may be reproduced or transmitted in any form or
by any means, electronic or mechanical, including
photocopy, recording, or any information storage
and retrieval system, without permission in
writing from the publisher.

Requests for permission to make copies of any
part of the work should be mailed to: Permissions,
Harcourt Brace Jovanovich, Publishers, 757 Third Avenue,
New York, N.Y. 10017.

Library of Congress Cataloging in Publication Data
Washburn, Mark.
Distant encounters.
1. Jupiter (Planet)—Popular works. 2. Saturn
(Planet)—Popular works. I. Title.
QB661.W37 1983   523.4′5   82-15709
ISBN 0-15-125744-2
ISBN 0-15-626108-1 (pbk.)

Designed by Joy Chu

Photographs by NASA/JPL

Printed in the United States of America

First edition

B C D E

THIS BOOK IS DEDICATED TO THE MEN AND WOMEN
OF THE JET PROPULSION LABORATORY AND OF
THE AMES RESEARCH CENTER, AND TO THE MEMORY
OF FELLOW VOYAGERS
EVERLY DRISCOLL, TIM MUTCH, AND NICK PANAGAKOS.

CONTENTS

Preface xi
Acknowledgments xiii

1. SPACE AGE 3
2. IN THE REALM OF GIANTS 16
3. THE MISSION 45
4. THE MESSAGE 69
5. FIRST ENCOUNTER 83
6. ATTACK OF THE SPACE GYPSIES 118
7. INSTANT SCIENCE 131
8. PIONEERING 162
9. NO SMALL RAPTURE 181
10. THE LONG GOODBYE 220

Index 263

Color illustrations appear between pages 176 and 177.

PREFACE

For years, I have had a recurring journalistic fantasy. By some process which does not bear close examination, I am whisked back in time to the year 1803. I track down Meriwether Lewis and William Clark and ask them if I can join them on their expedition; I want to write about their historic journey. Clark grumbles that they don't need some damned tenderfoot stumbling after them, but Lewis, a man of letters, says, "Sure, come on along!" And I do.

In the past three years, I have had the unexpected opportunity to live out that fantasy. I have stumbled along after a band of modern explorers on a fantastic trek to regions far more exotic and mysterious than even the untracked American West of 1803. I have had a front-row seat for one of the most awe-inspiring enterprises in all of history: the exploration of Jupiter and Saturn.

This book is a record of that odyssey, a twentieth-century journal of exploration and discovery. My goal is simply to take you, the

PREFACE

reader, along with me, to show you what I saw, to let you feel some of the excitement and anticipation I felt.

If I had actually traveled with Lewis and Clark, I suspect I would have written about mountains and rivers, Indian tribes and grizzly bears. I would have written more about the discoveries than about the discoverers, I suppose, because it is the object of exploration, rather than the process, that holds the greatest fascination for me.

So, too, with this account. This is not an "inside" story of the tools, procedures, and innermost thoughts of the explorers. Those subjects are certainly an important part of the story of the exploration of Jupiter and Saturn, but they are not the primary focus of this story.

The full story of Jupiter and Saturn remains to be told. The data collected by our spacecraft will take years to analyze and understand. What I present here are not the final judgments of the planetary scientists, but, rather, their first impressions. The "instant science" I describe here is only a small part of the full scientific enterprise.

But first impressions are important, too, for what they tell us about the discoveries and for what they tell us about ourselves. That first instant of discovery—whether it is a scientist's first view of the surface of another world or an infant's first step out the front door—is a unique and infinitely valuable human experience. In the moment of discovery, we are transformed; we become more than what we were only seconds earlier.

We can't all travel with Lewis and Clark, but every one of us can experience transcendent instants of revelation and discovery. Each day, we discover our world anew. We make so many discoveries that the experience becomes mundane, routine. But at rare moments in history, we have the opportunity to make discoveries of such scope and significance that they compel us to re-examine all that we know or think we know, and give ourselves over to the unbounded rapture and astonishment of Exploration. This is such a moment.

So, as Meriwether Lewis might have said, come on along! I hope you enjoy the journey as much as I did.

ACKNOWLEDGMENTS

This book is the product of three years of work, some 30,000 miles of travel, and more cigarettes than I care to think about. It is also the product of the efforts of hundreds of people I know and thousands more I have never met. I owe each of them a substantial debt of gratitude, but limitations of time, space, and memory make it impossible for me to thank each of them by name. Instead, I am forced to thank them by groups, en masse, and trust that the individuals involved will know that my gratitude is not merely collective, but specific and personal. And deeply felt.

First and foremost, I am grateful to the scientists, engineers, and managers of the Voyager and Pioneer missions at the Jet Propulsion Laboratory, the Ames Research Center, and NASA. Their energy, genius, and dedication are monumental; their achievements, historic.

To the men and women of the Public Information Offices of JPL, Ames, and NASA, my thanks is especially heartfelt. This book

ACKNOWLEDGMENTS

—indeed, any book or article on this subject by any author—would not have been possible without their enthusiastic cooperation.

My fellow denizens of the press room of the Jet Propulsion Laboratory also deserve thanks. In this book I have taken the liberty of referring to them (us) as a "motley group" of "space gypsies." I can already hear some of them objecting to this slander—in some cases, I imagine, at considerable volume. But (let's face it, folks), in all honesty, we *are* a motley group—and I am deeply proud to be a part of it.

And to my friends and family, I can only say thanks once again. They not only make it all possible, they make it all worthwhile.

Finally, a curious sort of thanks which requires a brief story. One afternoon during the Voyager 2 Saturn encounter, I was sitting at my desk in the JPL press room, doing whatever it is we journalists do when we are trying to look busy, when a young man came up and introduced himself. His name was Nicholas Booth, he was seventeen years old, and he was from England. He told me that he had read and enjoyed my book about the Viking mission, *Mars at Last!* That, in itself, was something of an event, since writers of modestly selling books seldom come into direct contact with their readers. But Nicholas went on to tell me the story of how he came to be at JPL. It seems that he had bombarded NASA with letters and inquiries about how he could arrange to be at JPL for the Saturn encounter, which would be the last such planetary encounter before he was well into his twenties. Through pluck and perseverance, Nick somehow managed to secure a job as a junior assistant in the JPL television operation. From that exalted position, he had a kind of worm's-eye view of the whole kaleidoscopic show, and he was clearly enjoying every minute of it. I admired Nick's enterprise, and, in a way, I envied him—when *I* was seventeen, I spent the summer selling housewares in a department store.

It would be very presumptuous of me to ascribe any cause-and-effect relationship to Nick's reading my book and his presence at JPL. His interest in space is deeply rooted (his fifth birthday came on the day Apollo 11 landed on the moon, his twelfth on the day Viking 1 touched down on Mars) and he plans to become a physicist; I am certain the day will come when I will be asking Dr. Booth to explain his latest data. But I do think, or hope, that there is some connection

ACKNOWLEDGMENTS

between books such as mine and the aspirations of young people such as Nicholas Booth. Writing is a tough business and the rewards are often meager; it is immensely gratifying to those of us on this side of the typewriter to know that our words do matter.

So, a final thank-you is in order. The story of the exploration of Jupiter and Saturn is the sort of heroic saga that once would have been told around a crackling campfire or before the hearth of a dockside pub. Tellers of such tales could see their listeners and know that their words had touched other hearts, other minds. I can't see *you*, gentle readers, but I know you're there. And I'm very grateful to you.

MARK WASHBURN
October 1982

DISTANT ENCOUNTERS

1 SPACE AGE

If the two Voyager spacecraft were persons with names and faces, they would be recorded in the history books on the same pages with Columbus, Magellan, and Lewis and Clark. But the Voyagers are simply mechanical contraptions, sophisticated and expensive, but ultimately just well-designed collections of nuts and bolts, devoid of blood and feeling and the visceral stuff of heroes. There will be no holidays, no statues in the park, no towns or universities bearing the name of Voyager.

It seems a shame. Because the Voyagers blazed a trail of exploration and discovery that is unmatched in all human history.

The scale of Voyager achievements is overwhelming. The two spacecraft revealed to our astonished eyes not simply continents and oceans, but entire worlds, vistas so broad and dazzling that it may be decades or even centuries before we fully comprehend all that Voyager showed us.

The Voyagers were launched from their home planet in August

and September of 1977; by September, 1981, they had completed their basic mission. In a voyage of comparable duration to that of Magellan, the spacecraft explored two planets, some twenty moons, two stunning planetary ring systems, and the complex interaction of fundamental physical forces in the vast gulfs between worlds.

After setting sail into uncharted seas, Magellan never again returned home. Neither will the Voyagers. Their work completed, the spacecraft will leave the solar system and roam forever through the silence of interstellar space. There in the galactic void, the Voyagers will likely survive the planet of their birth; Earth will be burned to a cinder in the death throes of the sun, ten billion years from now, but the Voyagers will endure, relics of the clever and curious species that created them. The holidays and statues, then, are not really necessary. The Voyagers will become their own memorials—and perhaps ours as well.

The Space Age is now a quarter of a century old—twenty-five tumultuous, and sometimes terrible years since Sputnik 1 swooped over our heads like some biblical omen, pulling our gaze skyward. In the years since 1957 we have taken our first tentative steps into the universe, scouting our immediate neighborhood and finding not just new worlds, but a new sense of ourselves. We are not the same as we were.

Much else has happened. Sputnik orbited a planet inhabited by three billion human beings grouped under the flags of some seventy nations. Today, the population has increased by about a third; the number of flags has more than doubled. Perhaps as many as a dozen nations have now stenciled their flags on weapons of mass destruction, and that number will also increase. In 1957, the survival of mankind depended upon the wisdom and forbearance of the leaders of two great nations; today, the same responsibility rests in many hands, not all of them sure or steady.

As we invested our wealth and genius in the creation of new means of deliberately killing ourselves, we discovered that we were already proficient in doing it unintentionally. The two treasures that make our planet unique in the solar system, our oceans and our atmosphere, were becoming repositories of poison. We had rendered parts of our planet uninhabitable, and we were belatedly realizing that what had already happened to a town, a river, or a lake might someday happen to an entire continent. Gazing back at ourselves from the moon, seeing our blue and gleaming planet as a whole for the first time, we realized that the Space Age is also the Earth Age. We began to take better care of our planet, or tried to.

Our space missions became metaphors. We were crew members, all of us, on Spaceship Earth. Like the astronauts in their cramped Geminis, Vostocks, and Apollos, we were dependent upon an awesomely complex but fundamentally fragile Life Support System. A space traveler who wasted fuel and fouled his own air and water would not survive for long.

We began to understand that our Blue Planet was part of a family of worlds. It was neither eternal nor changeless. The same forces that dictated the histories of our sister worlds were at work here as well, shaping our future as surely as they had determined our past. On Earth, as elsewhere in the solar system, planet-wide catastrophes, geologic upheavals, and dramatic alterations of climate had happened before and, with or without the intervention of man, would happen again.

If the planet and our perception of it have changed, so have we, ourselves, changed. In twenty-five years, Americans have seen one President assassinated, another President resign in disgrace, and three others driven from office. The patience and placidity of the fifties have vanished. In the midst of an unprecedented prosperity, we underwent sweeping changes in society and fought and lost a brutal war. Now, in the eighties, prosperity has given way to economic uncertainty, we feel ourselves hostage to inflation, crime, and terrorism; and polls reveal that for the first time the average American believes that his children will inherit, not a better world, but a worse one.

In the face of such a decline (real or perceived), in an era of conflict, fragmentation, and anxiety, the exploration of space may seem to be monumentally irrelevant. A mother in Newark who can no longer get food stamps cannot be expected to care very much about the discovery of active volcanoes on a moon of Jupiter. An analysis of Martian soil by a billion-dollar robot matters less to a Nebraska farmer than the condition of his own soil. To a fireman in Boston, an auto worker in Detroit, and a teacher in California, all of whom are about to be laid off, the surface temperature of Venus is not a high-priority concern.

People have also become suspicious of the political motivations behind the exploration of space. In the early days of the Space Age—when it was being advertised as a "space race"—few questioned the need to meet the challenge of Sputnik. The Russian presence in space was seen as no less of a threat than the Russian presence in Cuba. With the hammer and sickle flashing over our heads every ninety minutes, it seemed imperative to get Old Glory up there, too.

So Americans went to the moon. The Apollo Program was conceived in 1961 by President Kennedy and his advisers as a kind of shock therapy for a nation that was stunned by Soviet triumphs in space and American failure at the Bay of Pigs. American know-how, American pride, American courage would get American astronauts to the moon ahead of the Communists. The first moon landing would also have been a capstone for Camelot, a glorious climax to the second term that never was.

The road to the moon led through Dallas, the jungles and paddies of Southeast Asia, and the steaming streets of Watts and Detroit. By the time the great prize was won, it was already tarnished. When the American astronauts lifted off from the moon, they left behind a plaque that bore the name of, not John F. Kennedy, but Richard M. Nixon. The plaque proclaimed that we had come in peace for all mankind.

That Nixon's name was memorialized on the moon was, for many Americans, somehow symbolic of all that had gone wrong during the sixties. While American technology went peacefully to the moon for all mankind, it was also engaged in killing part of mankind in Southeast Asia and was failing to meet the desperate needs of America's own minorities. The gap between rhetoric and reality had grown too great for many to accept. For young Americans, in particular—the generation that had grown up with the Space Age and ought to have been its most enthusiastic advocates—the exploration of space came to be seen as just another gaudy sideshow in a carnival run by scoundrels.

The monumental achievement of sending men to the moon was somehow diminished by the way in which it had been done and the reasons for which it had been attempted. The whole enterprise was wrapped so tightly in the flag that the fundamental reasons for going were all but smothered.

By 1969, moreover, the original reason for launching the program seemed irrelevant. The Soviets weren't going to the moon, détente was in the air, and the strident rhetoric of the Cold War was—temporarily, at least—passé. If it was no longer necessary to beat the Russians to the moon, was it necessary to go there at all?

Without the boost of patriotic motives, the later Apollo missions were cancelled, and the last American left the moon barely three years after Neil Armstrong first stepped out onto the ancient dust of the Sea of Tranquility. The astronauts left behind several scientific instruments, but these were all turned off a few years later in a budget-cutting move.

By the mid-seventies, a nation wearied by war and Watergate seemed to turn its back on space exploration. NASA budgets were slashed; new programs were cancelled, and for six long years no American ventured into space.

Finally, in 1981, the Space Shuttle *Columbia* roared skyward and Americans were "back in the space business to stay," as Astronaut Robert Crippen put it. But the shuttle was a scaled-down, compromise vehicle which cost more and delivered less than it was supposed to. Only the Defense Department, which had been a prime force behind the funding and design of the shuttle system, had a certain use for the shuttle—as a military vehicle.

And back on Earth, the drums were once again beating. Twenty years after Alan Shepard's first flight—when the Soviets had troops in Cuba and troops poised ominously around Berlin and the Americans were beginning to send advisers to Vietnam—Young and Crippen flew the *Columbia* around a world in which the Soviets now had troops in Afghanistan and troops poised ominously around Poland and the Americans were beginning to send advisers to El Salvador. The eighties were starting out like a weird, negative-image replay of the sixties, complete with fevered ideologues, Cold War saber rattling, and an assassination attempt on an American President. We seemed destined to push the same rock back up the same hill.

Yet through it all, something curious was happening. It was a unique, unexpected, sometimes downright spooky phenomenon that flowed equally from fact and fantasy. There was in it a restless stirring of ancient dreams.

This strange new phenomenon was nothing less than the Space

Age, struggling to achieve consciousness and a voice. It was an infant, annoying, perplexing, misunderstood, easily ignored; but it was among us now, growing, learning, groping for an identity. Slowly, it began to make its presence felt.

The year 1968 was one of the loudest, noisiest years anyone could remember. It reverberated with the sounds of gunfire in Vietnam, Memphis, and Los Angeles, and the shouts of angry demonstrators from Chicago to Paris to Prague. Yet, in the midst of the cacophony, a brand-new note in the human symphony was being struck that year. Stanley Kubrick's film *2001: A Space Odyssey* opened to baffled, indignant reviews by critics who wondered what all those apes, light shows, and mystifying black monoliths were supposed to mean. Who would pay to see such an odd, incomprehensible movie? That same year, the NBC television network tried to cancel a science-fiction series which was getting mediocre ratings. But the network executives found that while other programs might have larger audiences, none had so fanatically *loyal* an audience as *Star Trek*. An avalanche of angry letters and telegrams forced the network into making an unprecedented on-air announcement that *Star Trek* had been renewed for another season.

That loud, painful year came to a close, really, on Christmas Eve. From orbit around the moon, that solemn night, came the voices of three brave men reading from the Book of Genesis. Something about that moment cut straight through the anguish and anger and touched us at the center of ourselves. It was a moment that seemed to hold out some bright but indistinct promise, vaguely glimpsed, of an age when poetry and progress could sail together in the same ship.

The infant Space Age had spoken its first words.

In due course, many of the sorrows, uncertainties, and failures of the past twenty-five years will be forgotten. What will be remembered is that this was the age when mankind first blazed a trail into the universe. In the face of this truly great achievement, much that overwhelms us now will melt away with the flow of time. Five centuries from now, perhaps our descendants will recall that while we were beginning our historic exploration of the solar system, we were also engaged in pointless wars, economic strife, and various social experiments. Perhaps they'll wonder if we understood and appreciated the importance of what we had begun. But, almost certainly, someone reading those yet-to-be-written histories will pause, try to imagine how it must have felt when mankind first stood on the moon or viewed the rings of Saturn, and then say to himself, "What an exciting time that must have been!"

If we went into space because the Russians were there, and if we went to the moon for political reasons, at least we went. And having gone, we found that, as the national mood changed, we could reduce or de-emphasize our commitment to space but we could never totally abandon it. In a sense, we were trapped by our own rhetoric.

Selling the moon program to Congress and the American public had required a certain amount of propaganda. But flag-waving alone wasn't enough; nor was it an appropriate point to emphasize to the rest of the world. One of the officially stated reasons was that we wanted to explore the universe and acquire knowledge that would benefit all mankind.

Our commitment to acquiring that knowledge was often thin. The early Mercury and Gemini astronauts were impatient with the scientific experiments they were required to perform; they were caught up in the challenge of precision flying. For the Apollo program, the moon landings were, above all, an engineering triumph; science had to take a back seat. And, of course, it *was* a race, and nobody expects a runner to drop to the track after breaking the tape and to begin doing chemical analyses of the cinders. In the end, we sent only one scientist-astronaut to the moon, on the last mission.

Apollo missions accomplished, we might have shut down the entire space program had the voice of the Space Age not already pierced our consciousness. NASA had already begun a program of unmanned planetary exploration in the early sixties. That, like Apollo, was at least partly a response to the Russians, who were the first to launch an interplanetary probe. Compared with the manned space-flight program, the planetary shots were extremely cheap, scientifically rewarding, and very good for the NASA image. The planetary program was pure science.

It was difficult to be cynical about planetary exploration. The politicians who approved the programs may have had other motives, but the people who designed, built, launched, flew, and analyzed the results of the planetary missions truly believed in what they were doing. They took seriously all the rhetoric about knowledge and exploration; their eyes really were on the stars.

The result of that dedication was spectacular success. We not only sent probes to the planets, but we did it incredibly well. From a strictly American perspective, exploring the planets may be the thing we do best. In the years since 1957 we have seen our national self-esteem pummeled by events beyond our control and by failures that we might have controlled. But when it comes to exploring the planets, we are still the best.

If our planetary probes did nothing more than remind us of our own ability and genius, they would be well worth the price. For all our problems, we remain a clever, talented, and inventive nation, capable of great accomplishments.

In 1957, after four billion planet-bound years, we earthlings took our first tentative steps into the cosmos. Our knowledge of what lay beyond our home planet was minimal. Five thousand years of astronomical observations had produced no more than a vague and incomplete picture of the solar system we shared with eight other planets. Twentieth-century astronomers didn't even pay much attention to the solar system, preferring to concentrate on the more distant phenomena of stars and galaxies. Although the planets were relatively nearby, there simply wasn't a lot that could be learned about them by observations from even the best telescopes. Earth's atmosphere blurred our view of the planets; until we could rise above it and make observations from space, our knowledge and understanding of the solar system were destined to remain severely limited.

We did know that the solar system is divided into two distinct regions, separated by a zone of unconsolidated debris known as the asteroid belt. The inner region is populated by four small, rocky planets, Mercury, Venus, Earth, and Mars. Mercury is a small, airless lump of rock close to the sun; very little was known about it. Venus is larger, nearly the same size as the Earth, and at times it passes within twenty-six million miles of us. But the surface of Venus is cloaked by a dense, cloudy atmosphere, the composition and temperature of which were unknown. One could spend a lifetime staring at the featureless clouds of Venus without learning a single thing about what lay beneath them.

Mars is a more promising subject for observation. Smaller than

Earth, it possesses a thin, generally clear atmosphere that permits an indistinct glimpse of its mysterious surface. Mars has polar caps, like the Earth, but it was impossible to determine by observation whether they are composed of frozen water or frozen carbon dioxide. The rest of Mars is mottled by light and dark patches. To the observer on Earth, the light patches appear rust-red, and the dark patches may or may not be greenish; the distorting effects of our atmosphere make even the determination of color a difficult business. If the dark areas were indeed green, it might just possibly mean that we were seeing Martian vegetation—life on another world. It was a fascinating speculation; the dark areas do change seasonally in both shape and size, rather like terrestrial vegetation. But it was known that, compared with Earth, Mars is cold and its atmosphere—composition unknown —extremely thin. Vegetation thus seemed improbable, and by 1957 most astronomers leaned toward various nonvegetative hypotheses to explain the behavior of the Martian surface.

Mars is also the scene of what seem to be long, linear features; they had come to be known as canals. By the beginning of the Space Age, almost no one believed that these were literal canals, as had been proposed by Percival Lowell at the beginning of the century. There was even some question as to whether the linear features existed at all, but it was impossible to be sure. Through a telescope, some observers could see the canals and some couldn't; and even if these curious features were real, no one had a very convincing explanation for them.

The four inner planets, despite their obvious differences, have much in common. All are small, dense, and close to the sun. An appropriately protected man might stand on the surface of any of them.

The outer solar system, beyond the asteroid belt, is an utterly different environment. Jupiter, Saturn, Uranus, and Neptune are all immense, gaseous worlds, very low in density, and apparently composed of materials like hydrogen, helium, ammonia, and methane, which are scarce in the inner solar system. Their visible atmospheres are strangely marked, and it was uncertain whether any sort of solid surface exists below them. These giant worlds possess a collection of planet-sized moons, some of them as large as Mercury. There are also some novel peculiarities, such as Jupiter's famed Great Red Spot and

Saturn's beautiful rings. Beyond Neptune lies tiny Pluto, discovered barely a quarter of a century earlier in 1930. It is so small and so far away that virtually nothing was known about it.

This was the solar system, as it was known in 1957. In a sense, we knew as much about it as Columbus knew about the Western Sea in 1492. We sat on the shore and stared at the distant horizon, speculating about the wonders that might lie beyond it. But until we actually launched ships into those uncharted waters, the truth would remain hidden. It was time to explore.

In February, 1961, the Soviets launched a spacecraft called Venera 1, destination Venus. Like a ship disappearing into a cloud bank on the horizon, Venera 1, its radio receiver dead, soon vanished forever. Seventeen months later, the Americans made their first attempt at planetary exploration, and it, too, was a failure. Mariner 1 veered off course soon after liftoff from Cape Canaveral, and like so many of those early rockets, it was destroyed by the range safety officer. Of the first two planetary expeditions, one sank without a trace and the other ran aground in the harbor.

Mariner 2 became Earth's first interplanetary success. After a flawless launch, the Mariner 2 spacecraft encountered Venus at a range of 35,000 kilometers on December 14, 1962. It returned data indicating that Venus had a high surface temperature and no magnetic field. In itself, the information was not astonishing, but it was knowledge that could never have been gained from the surface of the Earth. These first bits of data were like the initial bricks in an edifice that was soon to grow to an enormous size.

The Soviet Union was the first to launch a vehicle toward Mars. Like Venera 1, Mars 1 also lost contact with Earth a few months after launch. Two years later, the first American Mars mission, Mariner 3, was also a failure. But its sister ship, Mariner 4, gave us our first close look at the Red Planet in July, 1965. The twenty-two fuzzy television pictures returned by Mariner 4 as it sped past Mars revealed a planet pocked with great craters, a world depressingly similar to our own lifeless moon. Four years later, Mariners 6 and 7 gave us 202 additional pictures of the Martian surface, showing more craters, but no sign of canals, vegetation, or Martians.

Meanwhile, two Soviet spacecraft, Veneras 3 and 4, successfully landed on the surface of Venus. For the first time in history, human artifacts rested on the soil of another planet. Less than a decade after we first left our own world, we had touched another.

Throughout the sixties and early seventies, attention was still focused on the Apollo program and its series of manned moon landings. The world watched as astronauts gamboled about on the gray lunar plains, joking, singing, hitting seven irons, revving up their lunar dune buggies. Yet the sight of men on the moon quickly became mundane. The moon itself was partly to blame; it is a dull, colorless, dead world, hardly an appropriate prize for such an extravagant undertaking. The men who went there called the moon "fantastic," but they said precisely the same thing about the ungainly machines that took them there. It *was* fantastic, but for many people it was also boring.

The unmanned planetary exploration program was different. The planets were turning out to be far more exciting than anyone anticipated. The lack of Mars-bound astronauts removed some of the human drama that had been implicit in the moon program, but it also forced each one of us to *become* an astronaut. With no one on the scene to tell us how fantastic it was, we had to project ourselves into the planetary pictures and see the solar system with our own eyes, not someone else's. And what we saw was . . . well, *fantastic*.

Late in 1971, Mariner 9 went into orbit around Mars, and a new era of exploration dawned. Previous missions to the planets had been flybys, sending back quick snapshots of unfamiliar terrain. But Mariner 9 returned data for months and mapped the entire planet down to a scale of kilometers or less. We found that the early Mariners had misled us. Far from being a dull, moonlike world, Mars is a place of incredible diversity and harbors more than a few mysteries. There are no canals, but there are hundreds of tiny channels which suggest that, at some time in the distant past, Mars had been a warmer and wetter world, perhaps even a home for some sort of life. We also saw immense volcanoes, layered ice caps, and a colossal canyon system which, in the time-honored tradition, was named after its discoverer—Valles Marineris.

Mariner 9 was followed in 1976 by Vikings 1 and 2, which actually landed on the surface of the Red Planet. The sense of resonance with earlier explorers became even stronger as Mars turned into, not simply a planet, but a place. Early in the mission,

Viking Project Scientist Gerald Soffen exulted, "I now know how Lewis and Clark must have felt when they began exploring the West. It's fantastic!"

That word again. Fantastic. And yet, it truly was fantastic, and this time the world seemed to realize it. Those first pictures of the brick-red Martian terrain dominated the world's newspapers, and hundreds of millions of people stared and wondered how it would feel to stand there on that strangely familiar rock-littered plain.

The inner solar system was becoming a familiar place. We live in it and understood it tolerably well, and our expeditions to Mercury, Venus, and Mars had filled in some of the missing pieces of the picture. The outer solar system, however, remained to be explored. Even the nearest of the outer planets, Jupiter, is five times farther from the sun than is Earth—half a billion miles. Merely contemplating the outer worlds required an order-of-magnitude expansion of our consciousness. The size of things out there is huge, the distances between worlds immense, and the time scales involved in any missions to the frontier of the solar system make it necessary to think in terms of years and even decades.

Pioneers 10 and 11 blazed the first trails into the outer dark. They were like small, efficient scout ships, checking out the route for the main expedition. The two Voyagers would follow, flagship vessels for this new age of discovery. If all went well, Voyager 2, after encountering Jupiter and Saturn, would crown the expedition with a flyby of Uranus in 1986 and finally Neptune in 1989. Thirty-two years after the launch of Sputnik, and just twenty-seven years after the first planetary encounter at Venus, mankind would have explored in some detail every major body in the solar system, with the single exception of Pluto. Such an achievement is without parallel, almost beyond comprehension.

The people who were responsible for the exploration of the solar system were well aware of the historical scope of their enterprise. Astronomer David Morrison described 1979 as "the Jubilee Year of planetary exploration." In December, 1978, Pioneer Venus sent probes to the surface of that cloud-shrouded planet. In the months that followed, Voyagers 1 and 2 encountered Jupiter and its family of moons. And in September, Pioneer 11 gave us our first close look at Saturn. Reflecting on the meaning of it all, Morrison said, "It's as if Da Gama, Columbus, Magellan, and Cook all returned from their voyages within a period of a few months."

There was an important difference, however. NASA official Tom Young pointed out that in the old days people went down to the docks, waved goodbye to Columbus, and then went home to wait, sometimes for years. "We're a privileged group," said Young. "No longer is it only the explorers that go on the vehicle." It is in the nature of unmanned spacecraft that all of us can go along on the voyage. "All we have to do," said Young, "is to sit back, relax, and enjoy the thrill of exploring a new world. . . . We can all be explorers."

We can all be explorers. . . . That is what the Space Age has given us. It has rekindled dreams and joys that seemed to have escaped us forever. It reminded us that the tedious, frustrating, painful world we have inhabited these last twenty-five years is not all there is. Much more awaits us; there is always more to discover, whether a billion miles away or as close to us as the labyrinths of our own minds. Kubrick's monoliths were symbols of our own curiosity, forever leading us from the known to the unknown. We can all be explorers; we have all always been explorers. To be human is to be an explorer.

And we're just getting started. . . .

2 IN THE REALM OF GIANTS

They assault the senses. Everything about them challenges the rational mind. They exist on a scale far beyond human experience. Long before the Voyagers first revealed to us the riotous, multichromatic clouds of Jupiter and Saturn's magnificent symphony of rings, we knew that these were extraordinary worlds. We stared at them and, like ants gazing at eagles, tried to comprehend a universe that could contain such grandeur and greatness.

If an explorer from the other side of the galaxy were to happen upon our solar system, he might well report that this particular corner of space consisted of the sun, Jupiter, and debris. Ninety-eight percent of the mass in the solar system resides in the sun; about seventy percent of what's left is Jupiter, and about seventy percent of the remaining thirty percent of two percent is Saturn. Everything else, by comparison, is dust.

Jupiter, aptly named for the king of the gods, is the fifth planet from the sun, revolving around our star at a mean distance of

778,000,000 kilometers, or slightly less than half a billion miles. Since such figures are meaningless in everyday life, it is difficult to get a true sense of the scale of things. This is a common, recurring problem in any discussion of Jupiter and Saturn, and we shall encounter it again. Trivializing statistics may be distasteful to the purist; in this case it may serve a useful purpose. Jupiter is so far away from Earth that if it were possible to travel there in an automobile, driving day and night at a safe, legal 55 miles per hour, the trip would take a little more than 800 years. At the other extreme, light, traveling at 300,000 km per second, takes more than 43 minutes to go from the sun to Jupiter.

At that distance, 5.2 times farther from the sun than our own planet, Jupiter takes 11.86 terrestrial years to make a complete revolution. As seen from Earth, it is the second brightest planet in the sky, outshone only by Venus—that despite the fact that the sunlight reflected from Jupiter is about fifty times less intense than the light reflected from Venus. Closer to the sun, Venus is visible only in the hours around dusk and dawn; but for six months out of the year, Jupiter dominates the midnight skies. Moving at a slow, regal pace, shining with a dazzling white brilliance, Jupiter was obviously the king. The ancient astrologers who first tracked the motions of the heavens were wrong about nearly everything, but they were right about Jupiter.

Galileo was the first to see the true face of Jupiter, turning his homemade telescope toward it in January, 1610. Even when seen through a small telescope such as Galileo's (estimated to have had a magnifying power of about eight, comparable to a pair of cheap binoculars), Jupiter looks big. That impression is mainly due to the distinct flattening of Jupiter's poles, which makes the equatorial region look swollen and bloated. The equatorial diameter has been measured at 142,800 km, while the polar diameter is about 133,500 km, a flattening of about 6.5 percent.

The polar flattening is due to Jupiter's rapid rotation. The rotation period was first calculated in the 1660's by Giovanni Domenico Cassini, who came up with the surprisingly short period of 9 hours 56 minutes for the Jovian day. Later astronomers found slightly different answers to the same problem, and it was eventually realized that different parts of Jupiter rotate at different speeds. The rotation pe-

The Sizes of Earth, Jupiter, and Saturn

Drawn to scale, Earth looks insignificant compared with the gas giants.

riod typical of the equatorial regions is 9 hours, 50 minutes, 30 seconds.

It was possible to measure Jupiter's rotation because the planet is distinctly marked with dark bands and bright and dark spots which serve as temporary reference points. Galileo never saw the markings because his instrument wasn't powerful enough, but twenty years later a number of astronomers were sketching the bands and spots in their notebooks. The most prominent of these spots, known as the Great Red Spot, was probably first recorded in 1664 by Robert Hooke. Two Cassini drawings, dated 1672 and 1691, also show the spot.

It was not until 1878 that astronomers paid much attention to the Great Red Spot. German astronomer Wilhelm Tempel wrote a report about it that year, noting that the spot was about twenty degrees south of the Jovian equator, measured about thirty thousand miles long and seven to eight thousand miles wide, and was brick red in color. Tempel's "discovery" inspired other astronomers to check the old records, and they found a number of possible sightings dating back to Hooke and Cassini. The Cassini sightings were in some dispute, because his drawing of it looked distorted. Confirmation of Cassini's observation came from an unlikely source—the Vatican picture gallery. A series of paintings, presumed to be the work of Donati Crespi, Cassini's contemporary, show various astronomical scenes, including Jupiter, complete with its Red Spot.

The spot is elusive because it is changeable. Its color varies from dark red to pale pink, and its length varies by as much as fifty percent. Even more perplexing to astronomers is the fact that the spot drifts, wandering completely around the planet over the last century.

The ambling Red Spot was a confirmation to astronomers that when they looked at Jupiter, they were seeing, not the solid surface of a planet, but rather the turbulent cloud tops of its atmosphere. With cloud belts at different latitudes rotating at different speeds and lesser spots of various colors appearing and disappearing, it was clear that Jupiter was a different sort of world than rocky, fixed-face planets like Mars or the Earth.

Jupiter's mass was first calculated by Friedrich Wilhelm Bessel in 1842, using a complex formula involving rotation speed, polar flattening, and the orbits of Jupiter's known satellites. He concluded that Jupiter had a volume 1,250 times that of the Earth, but a mass just 388 times as great. Today, those figures have been refined, giving a volume of 1,317 times the Earth's, and a mass 318 times greater—roughly equal to 2×10^{24} tons. These figures mean that Jupiter's density is just 1.34 times that of water, compared with 5.6 for the Earth.

Interpreting these numbers has been a continuing problem. To nineteenth-century astronomers, they suggested that Jupiter, beneath its clouds, was a glowing, incandescent planet, perhaps similar to the young, molten Earth. Some proposed that the Great Red Spot was simply a reflection of some enormous, seething lava flow on the unseen surface, and that other atmospheric spots were evidence of more eruptions. When the spot was found to be moving, however, such theories had to be abandoned.

The problem was that up to the last few decades, astronomers were forced to work with a very limited amount of rather bizarre data. Like the blind men encountering their first elephant, they had to make logical deductions based on fragmentary and sometimes misleading information. They simply lacked the hard facts necessary to marry descriptive astronomy with theoretical astronomy. Getting those facts from the surface of the Earth was a difficult and frustrating business; it is not surprising that during the first half of this century, most astronomers turned to more manageable pursuits, such as the study of stars and galaxies. Planetary astronomy became something of a scientific backwater.

It was easier to study the stars because there were so many of them. Here in the solar system, we are stuck with just two basic kinds of planets: four small, dense planets, and four huge, light planets—plus (later) one extra planet so small and far away as to be almost impossible to study. But out in the galaxy, there are all sorts of stars, young, old, huge, tiny, bright, dim, exploding, solitary, double, multiple, clustered, veiled, pulsating, approaching, receding, white, blue, yellow, red, plus some odd phenomena which look like stars but proved not to be and clouds of gas and dust which do not look like stars but would eventually become stars.

Working with such a multiplicity of data, it was possible for scientists to evolve a broad theoretical framework capable of explaining the observed phenomena; when you're doing a connect-the-dots puzzle, it helps to have lots of dots. By the middle of this century, astronomers had a good (if broad and incomplete) understanding of what stars are, how they work, and how they evolve. This understanding, combined with the slow accumulation of data about our own solar system, allowed us to look once again at the planets and, for the first time, comprehend what we saw.

Based on evidence accumulated from meteorites, moon rocks, astronomical observations, and planetary encounters, the generally accepted theory states that 4.6 billion years ago, the solar system formed from a condensing cloud of dust and gas known as the solar nebula. The sun formed at the center of the nebula, its immense gravitational field pulling in most of the cloud's mass. Local areas of turbulence in the flat, spinning nebula created lesser gravitational whirlpools which consolidated the gas and dust into small clumps that eventually accreted into planet-sized bodies.

The planets all formed within the equatorial plane of the sun, also known as the plane of the ecliptic. In a sense, the solar system is two-dimensional, with virtually no matter present in the space above and below the equatorial plane.

In the inner solar system, the young proto-planets had to struggle for building materials with the sun. It was an unequal contest.

Heat and subatomic particles radiating from the new star drove most of the light, volatile gases out of the inner solar system, leaving behind four small planets composed of dense, molten iron cores and rocky surface materials. As the sun's intense radiation tapered off, gases trapped within the interiors of the planets gradually escaped from the rocks and formed the atmospheres of Venus, Earth, and Mars.

During the final stages of planetary formation, the solar system was swarming with leftover debris from the nebula. These unconsolidated pieces of space junk, some of them as large as a few hundred kilometers in diameter, smashed into the young planets, cratering their surfaces and triggering immense lava flows from their molten interiors. The scarred surfaces of Earth's moon and Mercury remain as silent witnesses to the incredible violence of our solar system's birth.

On Earth, the combination of time, air, water, and an active surface geology driven by heat from the interior has erased most of the ancient craters. On the moon, the cratering record is almost intact. The lunar craters have provided us a gauge with which to measure the relative ages of other planetary surfaces; where craters are numerous, the surface is old. Uncratered terrain implies youth and an active process (such as volcanism) which covers up or erodes the original craters.

In the outer solar system, the newborn planets were able to capture not just rocky material, but also the light, volatile gases in the solar nebula. Compounds such as water, methane (CH_4), and ammonia (NH_3), driven out of the hot inner solar system, became major constituents in the planet-building process. Today, evidence for the differences in inner and outer planet composition comes to us from meteorites, four-billion-year-old bits of cosmic flotsam and jetsam. Dense, heavy, nickel-iron meteorites seem to have formed in the inner solar system. Light, crumbly meteorites, known as carbonaceous chondrites, must have formed farther away from the sun, in the region of the giant outer worlds.

The chondrites are rich in organic molecules, composed of the basic building blocks of water, methane, and ammonia. Organic molecules are necessary for living organisms, and some scientists suspect that most of Earth's original life-giving organics probably arrived here in the form of carbonaceous chondrites. It seems likely that those same organics were abundant in the outer solar system, either as gases in the atmospheres of the giant planets, or as frozen ices locked into the bodies of the Jovian and Saturnian moons.

At some time not long after the formation of the planets, the sun apparently went through a particularly violent upheaval known as the T-Tauri stage. Streams of subatomic particles shot out from the sun and stripped the inner planets of whatever might have been left of their primordial atmospheres; Jupiter and Saturn—bigger, farther from the sun—managed to hold onto their gaseous envelopes. As debris from the nebula was being ejected from the solar system, the moons of Jupiter and Saturn must have experienced the same sort of bombardment that occurred in the inner solar system. Scientists expected that the moons of those worlds would look as dead and battered as our own much-abused satellite.

Although many of the details remain to be established, this version of our solar system's history seems to be in accord with what we know about how the universe operates. Our own solar nebula is long gone, but elsewhere (mainly in the arms of spiral galaxies) we can see other clouds of dust and gas. Embedded in many of those clouds are bright, hot young stars and, perhaps, unseen planetary systems in the process of birth.

More than half of the stars in our galaxy are double or multiple star systems, and for the sun, it was apparently a near-miss. Jupiter swept up most of the matter that did not go into the sun, but there wasn't quite enough left to turn Jupiter into a star. If it had been several times more massive than it is, the pressure and temperature at the center of Jupiter would possibly have been high enough to touch off the same hydrogen fusion reaction that powers the sun. The sun would have become a double star, creating an environment that might have made it impossible for life to evolve on Earth.

The composition of Jupiter is identical with that of the sun—and, in fact, with the composition of the observable universe. This solar composition consists of ten molecules of hydrogen for each atom of helium. All other elements exist in little more than trace abundances.

Even before the modern solar nebula theory was introduced, astronomers suspected that Jupiter must be rich in very light elements (hydrogen and helium are the two lightest) due to its low density. The first solid evidence for this came from a spectroscopic analysis of the light reflected from Jupiter, a technique which permits the identification of elements present in celestial bodies. Earth-based spectroscopy of the planets is difficult because our own atmosphere gets in the way, and because the planets shine by light reflected from the sun—thus, planetary spectra tend to be almost identical with the solar spectrum. But in 1905, V. M. Slipher detected several anomalous lines in the Jovian spectrum, which Rupert Wildt identified in 1932 as belonging to ammonia and methane. Ammonia (NH_3) and methane (CH_4) are hydrogen-rich compounds, and their presence in the Jovian atmosphere suggested that the bulk of Jupiter was composed of hydrogen.

Later, the molecular weight of Jupiter's atmosphere was calculated to be 3.3. The molecular weight of helium is 4, and of hydrogen (H_2), 2. It was obvious that the 3.3 figure represented a mix of hydrogen and helium.

Most current models of Jupiter and the other gas giant planets assume the presence of a molten, iron-silicate core at the center of the planet. The assumption rests on the "solar composition" argument. During its formation, Jupiter, like the other planets, must have received a proportionate share of the heavier elements that were present in the solar nebula. Since those elements have not been detected in the Jovian atmosphere (and one wouldn't expect to see much iron floating in an atmosphere anyway), it is reasonable to suppose that the heavier elements have congregated in a planetary core. Jupiter's core is thought to be no more than a few thousand kilometers in diameter, or about the size of the Earth.

Scientists now suspect that above the iron-silicate core of the planet, are two thick layers composed mainly of hydrogen. The bottom layer, calculated to have a radius of about 46,000 km, consists of liquid metallic hydrogen, a substance never seen on Earth. Its existence is inferred from what little we know about the behavior of materials under enormous pressure. As a metal, hydrogen ought to be an excellent conductor of electricity, lending some interesting and unusual properties to the Jovian interior; it may affect the way heat is transferred from Jupiter's interior to its atmosphere. This is the realm of almost pure theory, since observation and experimentation are virtually impossible in this case. No laboratory in the world could reproduce the calculated pressure at the level of the metallic hydrogen layer—estimated to be about 3 million bars, or three million times the atmospheric pressure at sea level on Earth.

The second layer of hydrogen, under less pressure from what lies above it, is thought to consist of liquid hydrogen in its normal molecular state. This layer extends outward to a distance of about 70,000 km from the center of the planet. Finally, the standard model proposes that above the liquid hydrogen is a 1,000-km-thick atmosphere composed of still more hydrogen, now in its gaseous state.

The present model of the Jovian interior evolved slowly, and will undoubtedly be subject to further modification. Whatever the details, the central point is that beneath the striped beachball atmosphere of Jupiter is nothing resembling a solid surface. The very idea is something of an affront to our inner solar system sensibilities, but it

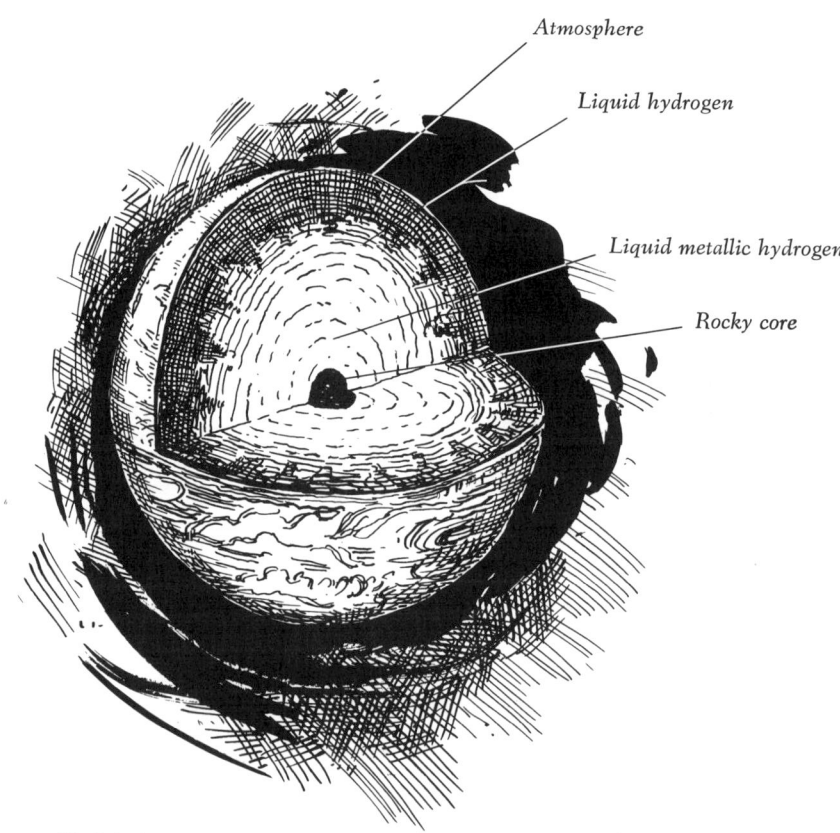

Model of Interior of Jupiter

In the standard pre-Voyager model of Jupiter, the only solid portion of the planet is the rocky core. The rest of it is mostly hydrogen in liquid and gaseous states.

is a reminder that the outer solar system is an environment utterly different from the one we know. One can attempt to understand Mars or Venus by using Earth analogies, but such exercises may be seriously misleading when applied to the gas giants.

One Earth analogy, popular in the nineteenth and early twentieth centuries, led to the presumption that Jupiter was a hot, primitive planet, similar to the young Earth. Astronomers had observed seething turbulence in the colored bands of Jupiter's atmosphere. Atmospheric circulation requires the injection of energy. On Earth, the necessary energy comes from solar heating. The sun heats the tropical oceans, which then transfer their energy to the atmosphere, driving the global circulation system. But Jupiter is so far from the sun that it receives only about four percent as much solar energy per unit area as does the Earth. It was thus assumed that the energy in the Jovian atmosphere must come from an internal source, most probably from heating due to the natural radioactivity of the heavy elements present in what was then thought to be the solid body of the planet. Jupiter was hot.

This picture was turned on its head in the 1920's and 30's. British scientist Dr. Harold Jeffreys calculated the amount of radioactivity that would be needed to produce the observed atmospheric turbulence. His answer: about 10,000 times more than is present in the Earth. That was too much, even for Jupiter. Meanwhile, Harvard astronomer Donald Menzel had taken measurements that indicated a temperature at the top of the Jovian atmosphere of about −110°C. When Wildt identified ammonia and methane—compounds that remain gaseous even at very low temperatures—in the atmosphere, the picture seemed complete. Jupiter was cold.

Several models of the Jovian interior proceeding from the cold Jupiter picture included layers of ice somewhere between the hot core and the very cold upper atmosphere. Wildt postulated a mantle of water ice some 27,000 km thick. The ice would be under incredible pressure, forcing it into greater than normal density and an unusual structure—rather like Vonnegut's "ice-nine." Wildt's model was challenged by W. H. Ramsey, who did away with the ice and introduced a layer of solid metallic hydrogen. Both models survived into the 1960's, and both implied a cold Jupiter.

Before the twentieth century, scientific disciplines resembled island communities. There was a sporadic ferry service running between some of the islands, and occasionally a new idea from the chemists' island might gain currency among the residents of the biologists' island, or a discovery on the geologists' isle might be incorporated into the theories of the insular astronomers. For the most part, however, each island went its own way and paid no more attention to the others than was absolutely necessary.

Bridge-building began in earnest in the postwar years, as a new generation of scientists discovered that the resources of any one island were simply inadequate to explain all that needed explaining. The biologists and chemists built a bridge, and the new discipline of biochemistry was born. The geologists got together with the physicists and created geophysics. It was all rather like a case study in the benefits of cross-fertilization, and hybrid vigor asserted itself in a profusion of new, cross-disciplines such as astrophysics, geochemistry, planetary geology, archeoastronomy, biophysics, and paleoclimatology. The influx of new ideas also inspired the scientists to take a closer look at their own islands and investigate isolated coves and inlets that had previously been unexplored or inaccessible, resulting in entirely new fields of study such as radio astronomy and molecular biology.

By the 1950's, these new disciplines were beginning to transform our understanding of Jupiter and, in fact, the rest of the universe. Traditional astronomers had assiduously examined the visible wavelengths of light emitted from celestial objects, but visible light constitutes only a small fraction of the entire electromagnetic spectrum. Other wavelengths can provide additional information.

Everything in the universe is vibrating, and each type of atom releases energy in the form of waves traveling at the speed of light. The length of the waves is dictated by the amount of energy involved. Very long wavelengths, such as in microwave and radio emissions, carry relatively little energy; as the wavelength decreases—through infrared, visible light, and ultraviolet light—the energy level increases. At the short, highly energetic end of the spectrum, emissions are observed in the form of x-rays and gamma rays. Because the Earth's atmosphere absorbs energy at many wavelengths, not all parts of the electromagnetic spectrum are accessible to ground-based observers. But radio waves do get through, and the study of them has opened a new window on the universe.

In 1955, B. F. Burke and F. L. Franklin of Carnegie Institute detected a strong, irregular signal at radio wavelengths. They tracked down the origin of the signal and found, to their considerable surprise, that the source was the planet Jupiter. Other radio astronomers tuned in to the giant planet and picked up more Jovian broadcasts, some of them apparently synchronized with the planet's rotation. Some scientists thought the emissions were similar to a beacon, or the light from a lighthouse, spinning with the planet, shining in our direction once every revolution. The strength of these emissions was much greater than ought to have been the case for a planet as cold as Jupiter was assumed to be. The signals resembled interference from thunderstorms (which is what they were originally thought to be) and suggested that electrical phenomena similar to thunderstorms might be present in the Jovian atmosphere.

The mysterious radio emissions became less mysterious in 1959 when an instrument aboard Explorer 1, the first American satellite, detected evidence of belts of radiation surrounding the Earth. Known as Van Allen belts, after James Van Allen of the University of Iowa whose Geiger-counter experiment on Explorer 1 discovered them, these radiation belts are byproducts of Earth's magnetic field. The magnetic field lines of force extend out into space, where they trap charged subatomic particles such as protons and electrons (think of the particles as iron filings trapped in the field of a simple, rotating magnet). The area of influence of the magnetic field is known as the magnetosphere. Basic physics suggested that the magnetosphere could be a source of some of the radio emissions.

The intense radio emissions from Jupiter indicated that Jupiter had its own magnetosphere. Soviet astronomer I. S. Shklovsky noted that the emissions were similar to the intense radiation observed in high-speed electron accelerators on Earth. In the accelerators, electrons are magnetically whipped to velocities near that of light. Apparently the Jovian magnetic field had much the same effect, causing high-speed electrons to spiral along its magnetic lines of force.

The strength and size of the Jovian magnetosphere are enormous. One scientist later described the Jovian field as "the largest structure in the solar system." If it were visible to the naked eye, it would look larger in the sky than the sun.

The long-wavelength radio emissions gave us our first hint that Jupiter is something more than just a cold ball of gas. Slightly shorter infrared wavelengths provided another surprise. The water vapor in Earth's atmosphere makes it almost impossible to observe infrared emissions from celestial sources, since it absorbs the infrared wavelengths. Infrared astronomy didn't become a practical pursuit until the 1960's, when it was possible to send infrared detectors aloft on rockets and in high-altitude jets.

In 1969, one such detector was placed aboard a NASA Lear Jet, which flew above ninety-nine percent of the atmospheric water vapor. The experiment, run by Frank Low and his colleagues at the University of Arizona, produced a result that was nothing less than astonishing. Jupiter, that cold ball of gas, was actually emitting twice as much heat as it received from the sun. The cold Jupiter model went out the window. Jupiter is hot, after all.

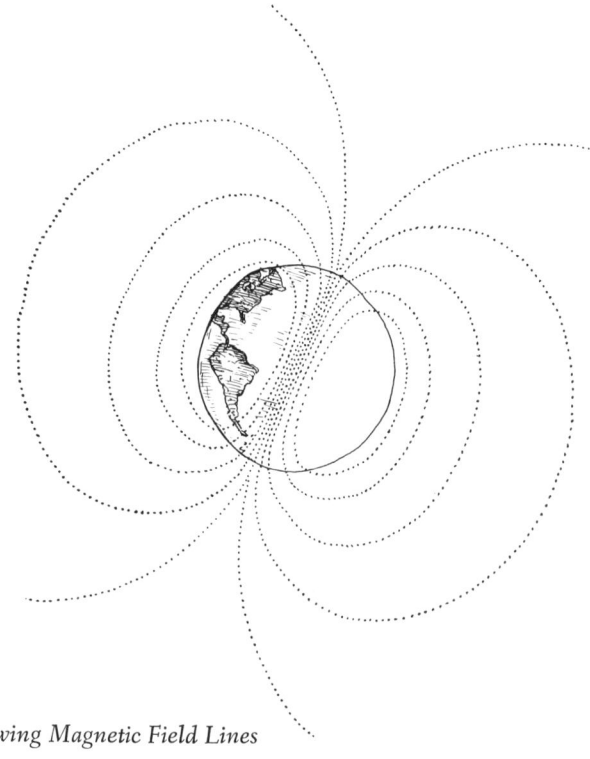

Earth Showing Magnetic Field Lines

The pattern of a planet's magnetic field is similar to that of iron filings around a magnet.

The earlier astronomers, as sometimes happens in science, had been right for the wrong reasons. Jupiter scarcely resembles the proto-Earth, and it has no great lava flows or vast radioactive heat sources. What it does have, it now appears, is a huge reservoir of heat left over from the original solar nebula. The heat emitted from Jupiter is, in effect, fossil heat, trapped for four and a half billion years in the high-pressure interior of the planet. As the planet slowly cools, the primordial heat is released and observed as infrared emissions.

At still shorter, more energetic infrared wavelengths, there were more surprises. Astronomers discovered Jovian "hot spots" at a wavelength of 5 micrometers. The hot spots seemed to radiate from holes in the upper-level cloud layers of the Jovian atmosphere. Beneath the frigid cloud tops, the atmosphere is much warmer. For each kilometer of descent into the atmosphere, the average temperature goes up by about 1.9°C.

A hot Jupiter made the turbulent atmosphere somewhat easier to understand. Solar heat could not provide the energy needed to run the atmospheric circulation system, but Jupiter provided its own heat. The alternating white and colored bands that give Jupiter its distinctive appearance seemed to be convection cells, or columns of heated, rising gas, stretched into horizontal streaks by Jupiter's rapid rotation. In the white bands, gas rises until it cools enough for the ammonia in the largely hydrogen atmosphere to freeze. As the cooling continues, the ammonia cloud in the atmosphere spreads in a north-south direction and descends. At the boundary with the next band, it is swept back into an east-west flow. By convention, the light bands, whitened by frozen ammonia, are known as zones, the dark bands as belts. The zones are higher than the belts; in the lower, warmer belts the ammonia remains in its colorless gaseous state.

The belts and zones have been charted for centuries and, from the Earth, appear to be well-defined, permanent structures in the Jovian atmosphere. The view from Voyager was to be quite different, and would necessitate a rethinking of the entire problem of the motions of Jupiter's atmosphere.

One of the enduring questions concerning the atmosphere is why there should be any colors at all in the belts and zones. The dark belts appear in many different shades, ranging from brown to red to yellowish to blue. Since the constituent gases of the atmosphere are

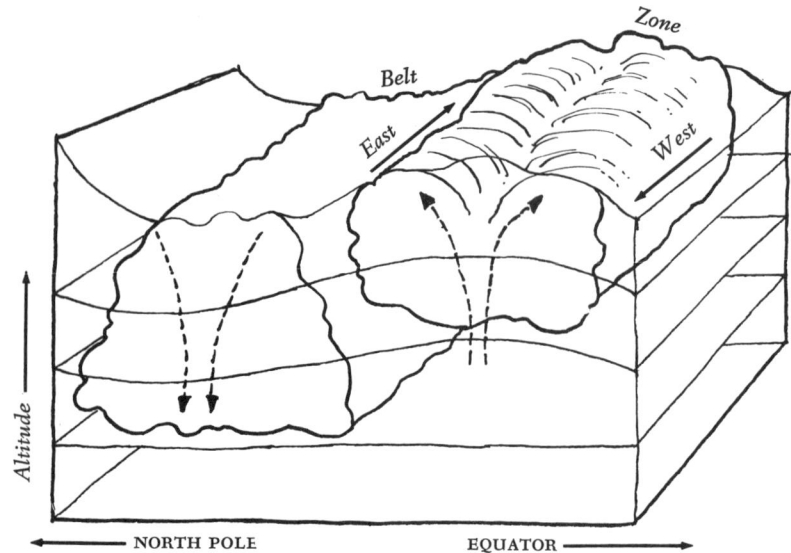

Before Voyager, scientists thought that the alternating east-west flows in Jupiter's atmosphere were created by hot, rising gases in the zones and cooler, descending gases in the belts. Voyager results suggest that what really happens is much more complex.

Model of Jupiter's Belts and Zones

colorless, it is obvious that we were seeing impurities in the gases. Somehow, material from below is being selectively injected into the belts, giving them their color.

Then there is still the Great Red Spot. Why is it red? And why has it endured for centuries, while smaller spots have come and gone on a time scale ranging from months to decades? By the beginning of the 1960's, one of the prevailing spot theories was that the Great Red Spot was an immense chunk of ice floating in the Jovian sky. Why or how that should have been the case was unclear, and ice on earth is not red. However, as the cold Jupiter was replaced by the hot Jupiter, the floating iceberg theory melted. The Great Red Spot was now thought to be the Jovian equivalent of a hurricane and the lesser spots seemed to be smaller storm systems. But hurricanes on Earth are not red either, or any other color in particular, so the redness of the spot may have been misleading or irrelevant. If it was truly a storm, it was a peculiarly Jovian storm.

Some fascinating possibilities exist. While Earth-based infrared observations remain a difficult undertaking, throughout the seventies astronomers succeeded in identifying a number of compounds in the top of the Jovian atmosphere. Among them: ethane (C_2H_6), acetylene (C_2H_2), water vapor, phosphine (PH_3), hydrogen cyanide (HCN), and carbon monoxide (CO). Most of these compounds are present in very small amounts—at the level of parts per million or parts per billion by volume—and none of them could be responsible for the atmospheric colors. But their very presence implied to scientists that some interesting and possibly exotic chemical reactions were taking place in the deeper Jovian clouds.

Although many experts believed that the cloud colors were probably produced by elemental sulfur, which can take on a variety of hues under differing conditions, others made the daring suggestion that the Jovian colors might be caused by complex organic molecules or even by living organisms. The gases in the Jovian atmosphere are the same as the gases believed to have been present in the atmosphere of the young Earth. In a classic experiment first performed in the early fifties by Stanley Miller and Harold Urey, it was discovered that when a combination of hydrogen, water vapor, ammonia, and methane is exposed to an energy input (in this case, an electric spark), it produces a number of hydrocarbon compounds, generically known as organic compounds, including some that are closely linked to protein molecules. Moreover, the organic sludge that accumulated in the bottom of the test tube had a reddish brown color reminiscent of the colors in the atmosphere of Jupiter. If the combination of gases and energy gave rise to the first living organisms on Earth, scientists wondered, could the same thing be happening in the clouds of Jupiter?

This seemed, to some scientists, to be an entirely plausible explanation for the coloring agents (called chromophores, literally "color carriers") in the Jovian clouds. All the raw materials were present; the existence of ethane and acetylene implied that some sort of hydrocarbon chemistry was going on, and deep in the atmosphere between the frozen cloud tops and the hot interior there had to be some level at which the temperature was biologically comfortable. The Jovian microbes might ride the wind currents, rising in the upwelling zones, falling in the deeper belts, somehow maintaining enough altitude to keep from frying in the hot, lower regions.

Cornell astronomer and exobiology advocate Carl Sagan went so far as to suggest a possible airborne Jovian ecology, made up of a population of "hunters, sinkers, and floaters." These organisms, which might resemble Portuguese men-of-war, would bob up and down in the atmosphere, regulating their own bouyancy to stay within a comfortable temperature range. That he made the suggestion does not necessarily mean that Sagan believes it is true, but the basic idea is well within the bounds of what might be possible.

In the years before Voyager, scientists debated organic vs. nonorganic explanations for the colorful Jovian clouds. Some possibilities were excluded, others were supported; no firm conclusions were reached, but the debate itself was fascinating. For decades, it had been an article of faith that if life existed elsewhere in the solar system, it would most likely be found on Mars, the most Earthlike of the other planets. But after the 1976 Viking mission failed to return a positive indication of Martian life (although the results were tantalizingly ambiguous), attention turned to Jupiter and the other gas giants. As NASA scientist R. D. MacElroy put it, "Living existence anywhere is becoming an acceptable phenomenon." Little Green Men need not apply, but other sorts of extraterrestrial organisms no longer seemed so preposterous. "Investigation of life in extreme environments," wrote MacElroy, "becomes a fascinating game, which may lead to some universal truths in biology, or possibly to more complete chaos in this captivating and often unruly science." MacElroy believed that life on Jupiter or Saturn was possible, "at least to those of us who willingly grasp at straws."

There were certainly some interesting Jovian straws to grasp at. A scientist such as UCLA's Willard F. Libby could speculate that "it seems likely that something like crude oil and proteins can be raining out on Jupiter"; but another scientist, Richard S. Hanson of the University of Wisconsin, could reply with equal justification, "The data available to me indicate that terrestrial forms of life could not have evolved or survived on the giant planets in this solar system." Whatever the truth of the matter, it was not likely to be discovered by Earth-based observations.

Whether the Jovian clouds were teeming with odd little beasties browsing and preying on one another or merely polluted with specks of sulfur, the truly strange thing about them was their persistence. As atmospheric expert Andrew Ingersoll of the California Institute of Technology explains it, "Instead of one westward current at low latitudes (the trade winds) and one eastward current at high mid-latitudes (the jet stream) as on Earth, Jupiter has five or six of each kind of current in each hemisphere." These zonal jets, as they are called, are located at boundaries between belts and zones, have velocities of over two hundred miles per hour, and have maintained their positions for at least several decades and probably for much longer. The dynamics of such an atmosphere seem incredibly complicated; in fact, by the time Voyager arrived at Jupiter, atmospheric scientists were coming to the view that Jupiter's atmosphere didn't resemble Earth's atmosphere so much as it did Earth's *oceans*. Thus, another bridge: this one, an unlikely structure connecting the wildly disparate disciplines of astronomy and oceanography.

For more than three centuries, Jupiter had exerted its insistent pull on the scientists of Earth. At first, it clutched only the astronomers, men like Hooke and Cassini, who sketched the great planet in their notebooks without having any real notion of what it was they were seeing. But on the eve of Voyager, it had firmly grasped the attention of cosmologists, physicists, chemists, meteorologists, oceanographers, geologists, and even biologists. Getting Jupiter to give up its secrets would, in the end, be a collective enterprise, bringing together men and women who had nothing in common except human curiosity and a sense of wonder. Especially the latter—it goes with the territory. The more we learn about Jupiter, the more awe-inspiring it becomes. From the king of the gods, we should have expected no less.

Saturn . . . the Jewel of the Solar System. Astronomers search the sky for knowledge and at Saturn they find, instead, beauty. The most distant planet known to the ancients, the cool golden gleam of Saturn stands like a sentinel at the gates to our galaxy, a billion miles from the fires of the sun. If Jupiter is the angry surf at Big Sur, then Saturn is Crater Lake or Walden Pond, a scene for quiet contemplation, beautiful not for what it does but for what it is.

What it is, fundamentally, is another immense ball of hydrogen and helium. Twice as far from the sun as Jupiter, Saturn takes 29.46 years to complete an orbit. Like Jupiter, it spins rapidly (10 hours, 39 minutes, 24 seconds) and is even more noticeably flattened at the poles; the equatorial radius is 60,330 km, the polar radius only 54,000 km. Saturn's volume is 815 times greater than Earth's, but its mass is just 95.2 times greater. Its density is so low (about half as

DISTANT ENCOUNTERS

dense as Jupiter and one-eighth as dense as the Earth) that Saturn would actually float in water. Vital statistics aside, Saturn is—to astronomer and nonastronomer alike—the most readily identifiable object in the heavens: it's "the one with the rings."

When Galileo first observed Saturn through his telescope in 1610, he thought that the rings were two very large satellites close to the planet. He wrote that they looked "like two servants supporting an old gentleman." Two years later, Galileo found that the servants had vanished. "What is to be said about so strange a metamorphosis?" he wondered. "Are the two lesser stars consumed after the manner of the solar spots? . . . Has Saturn perhaps devoured his own children?"

Galileo didn't know the answers, and for nearly fifty years, neither did anyone else. In 1659, Dutch astronomer Christiaan-Huygens, working with a much better telescope (of his own design and construction), concluded that Saturn was "encircled by a ring, thin and

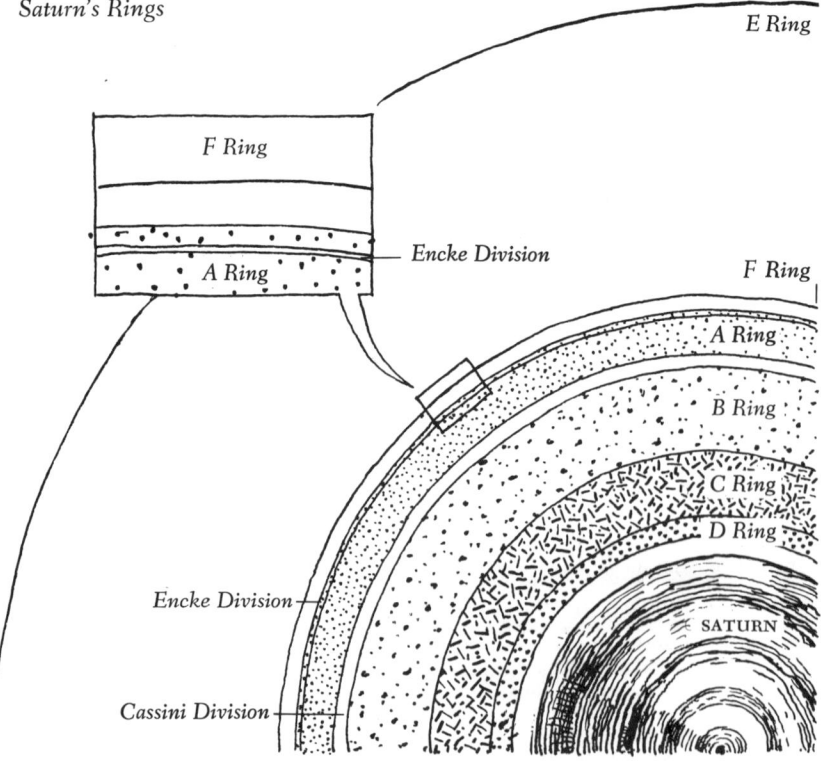

Viewed from Earth-based telescopes, Saturn's rings appear as distinct entities with gaps between. They are initialed in order of discovery, not position.

34

flat, nowhere touching, inclined to the ecliptic." Huygens' theory agreed with observation and explained why Galileo's "servants" had disappeared so mysteriously. Because Saturn was inclined to the ecliptic (the equatorial plane of the solar system), the rings would slowly bob up and down with respect to an observer on Earth. When the rings presented themselves edge-on, they disappeared from view altogether, implying that they were extremely thin. Huygens himself didn't believe they were *that* thin, and thought that the periodic disappearance was due to our seeing the dark, unilluminated side of the rings.

In 1675, Cassini made note of a "dark line" which separated the bright inner part of the ring from the slightly dimmer outer ring. This gap, which came to be known as the Cassini Division, was the first of several such gaps to be discovered; thus, the ring became "rings." The rings are identified by letters, assigned in order of discovery. By the time Voyager 1 reached Saturn in 1980, astronomers thought they had seen at least six major rings, A through F.

The outermost ring, the E ring, is very dim and not easily seen from Earth. It extends some 480,000 km above the cloud tops of Saturn. Next in, the thin and mysterious F ring was discovered by Pioneer 11 in 1979. Continuing inward, the A and B rings, separated by Cassini's Division, are the largest and most prominent of the rings. Inside the B ring, just 17,000 km above the planet, is the broad but dim C ring. And inside that a few astronomers thought they saw yet another ring, labeled the D ring. Since most astronomers never saw it, its existence was open to debate.

Identifying the rings was much easier than explaining their existence. Cassini suspected—correctly—that the seemingly solid rings were actually composed of swarms of small satellites, but he couldn't prove it. William Herschel, the great British astronomer, thought that each ring was a solid body; for a time, he even doubted that Cassini's Division was a real gap. The idea that the rings were solid was supported by the visual evidence, but as astronomers became more sophisticated in mathematics and physics, the solid rings started to crumble. It was clear that different portions of the rings would have to rotate around the planet at different speeds (the outer edge moving faster than the inner) and this would produce unendurable stress on any solid body.

DISTANT ENCOUNTERS

In 1857, the Scottish physicist James Clerk Maxwell produced a classic paper, *Essay on the Stability of Saturn's Rings*. In it, he proved mathematically that even if it were possible for the rings to be solid, if they were they wouldn't look the way they do. After disposing of the possibility of fluid rings, he concluded that "the only system of rings which can exist is one composed of an indefinite number of unconnected particles, revolving round the planet with different velocities according to their respective distances." Maxwell didn't calculate the size of the individual particles, but later observations implied that they averaged between one and thirty centimeters in diameter. Both larger and smaller particles were probably present. Spectroscopic analysis provided evidence that water ice was present in the ring particles; they may have been pure iceballs, or small rocks coated with a thin layer of frost.

The structure of the rings posed more questions. Why were there gaps in the rings? The standard theory proposed that the gaps were produced by gravitational interractions with the moons of Saturn. A particle orbiting in the Cassini Division, for example, would make a complete trip around the planet in 11.3 hours. But that orbital period is almost exactly half the period of the moon Mimas, a third the period of Enceladus, a quarter that of Tethys, and a sixth the period of Dione. This means that on every second orbit, the particle would close to minimum distance to Mimas and receive a small gravitational tug from the moon. On every third orbit, the same thing would happen with Enceladus, and so on. The overall effect of

1. *A ring particle in the Cassini Division arrives at Point a every 11 hours. The moon Mimas arrives at Point b every 22 hours.*

2. *On every second orbit, the ring particle aligns with Mimas at minimum distance and receives a small gravitational tug, pulling it toward Mimas.*

3. *The ring particle receives a similar tug from Enceladus every third orbit, from Tethys every fourth orbit, and from Dione every sixth orbit. These "gravitational resonances" remove most ring particles from the region.*

How Moons "Sweep Out" a Portion of a Ring

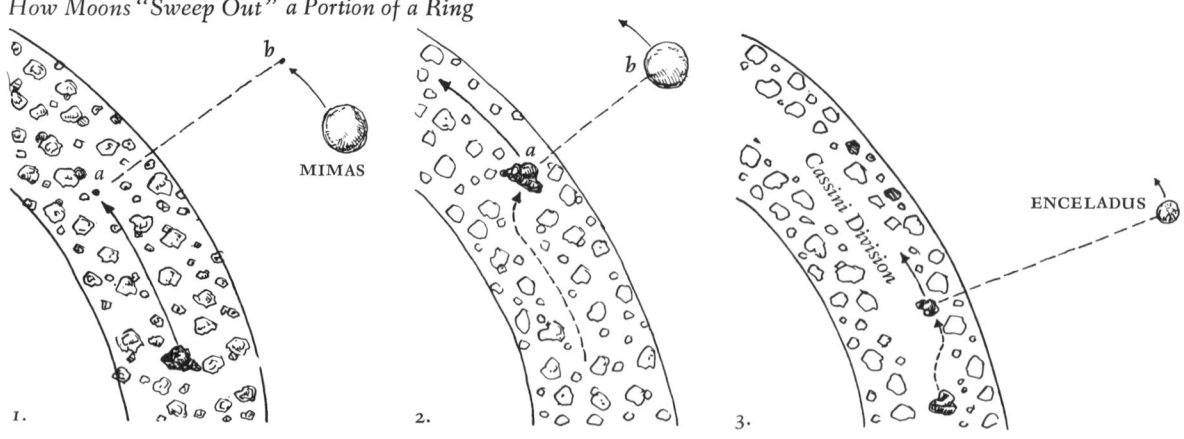

these gravitational resonances, as they are called, would be to remove particles from the region of the rings which orbited with a period of 11.3 hours, thus forming the empty Cassini Division. Other gaps would be created at other points in the rings where similar resonances existed. The gravitational resonance theory seemed to work fairly well as a broad-gauged catchall explanation for the ring observations. However, all the observations were made from Earth; the view from space was to be shockingly different.

Although the rings received most of the attention for many years, astronomers also studied Saturn itself. From the great size and low density of the planet, it was clear that Saturn was somehow similar to Jupiter; however, the visible cloud tops of Saturn were consistently bland and featureless, unlike the chaotic Jovian atmosphere. In 1876, American astronomer Asaph Hall, who later discovered the two tiny moons of Mars, observed a bright spot in the equatorial region of Saturn. The spot persisted for sixty-one revolutions of the planet and, although it didn't reveal much about the nature of Saturn, it was useful in refining the figure for Saturn's period of rotation.

In all, Earth-based observers have seen fewer than a dozen such spots on Saturn. If the spots are similar in nature to those on Jupiter, then the Saturnian spot-production process must be less efficient than that of Jupiter. Similarly, Jovian-style horizontal bands are also present in the Saturnian atmosphere, but there are fewer of them and their contrast is much lower than that between the white zones and colored belts in Jupiter's atmosphere.

Farther from the sun, Saturn's cloud tops are some 30°C colder than Jupiter's. At this lower temperature, the volatile gases such as methane and ammonia should freeze at an altitude well below the top of the atmosphere. This means that there could be a sort of hydrogen haze layer above the cloud tops, giving a filtered view of the atmospheric banding, resulting in the planet's muted appearance.

The composition of Saturn's atmosphere seems to be much the same as that of Jupiter. Methane and ammonia have been detected in small amounts, and other gases such as water vapor and hydrogen sulfide are assumed to be present, although the lower freeze level beneath the hydrogen haze has so far made them impossible to detect. But the main constituents of Saturn are hydrogen and helium.

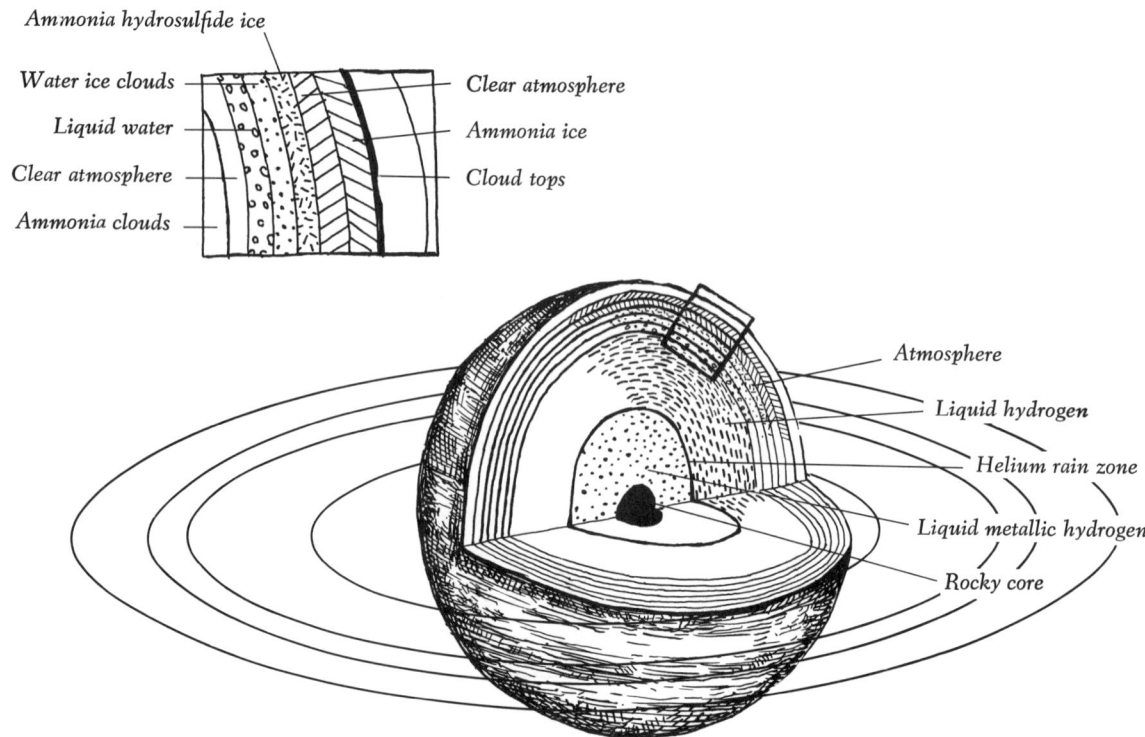

Model of Interior of Saturn

Hydrogen makes up about ninety percent of the molecular composition of Jupiter's atmosphere; for Saturn, the total is about ninety-four percent. The difference is significant. According to the "solar composition" argument, Saturn's atmosphere ought to have roughly the same hydrogen-to-helium ratio as Jupiter and the sun. But instead of a 9- or 10-to-1 ratio, the hydrogen-helium ratio on Saturn is about 16 to 1. Somehow, the helium in Saturn's atmosphere has been depleted.

The internal structure of the two planets is thought to be similar but not identical. On Saturn, the zone of metallic hydrogen should be smaller, due to the lower mass and internal pressure. On Jupiter, helium should mix uniformly with the hydrogen. But on Saturn, with a lower internal temperature, the helium is thought to be condensing to liquid form at the boundary between the metallic hydrogen layer and the hydrogen-helium envelope. As atmospheric scientist Andrew Ingersoll puts it, "raindrops of helium should be falling toward the center of Saturn." This process would account for the helium deple-

There is a possibility that helium rain is precipitating within Saturn's interior. The hypothetical structure of the atmosphere is shown here in some detail.

tion in the upper atmosphere. It would also explain another difference between Jupiter and Saturn. Both planets radiate more heat than they receive from the sun, but unlike Jupiter, Saturn is not large enough to have retained its original heat of formation; Saturn's modern-day heat source must be different from Jupiter's. Helium rain could provide for the Saturnian heat source; the gravitational energy of the falling helium would be converted to heat. The helium depletion process would have begun as soon as the internal temperature dropped to the condensation point of helium. On Saturn, that should have happened some two billion years ago. On Jupiter, the helium rain may be just beginning.

Perhaps the most striking similarity between Jupiter and Saturn is that each giant planet is the center of a miniature solar system. Before the Voyager encounters, Jupiter was known to have at least thirteen moons, Saturn ten. The existence of so many satellites is another indication that the outer solar system is a radically different environment. In the inner solar system, Mercury and Venus are moonless, Earth has its one large satellite, and Mars is circled by two tiny chunks of rock. The history of the outer solar system must have been very different for so many moons to have formed there. In addition to the Jovian and Saturnian systems, Uranus has at least five medium-size moons and distant Neptune is known to have two, including one that is larger than the Earth's moon. Even tiny Pluto has at least one moon, discovered in 1978.

The fact that the gas giants possess so many satellites suggests that as they were forming, they mimicked the formation of the entire soon as he looked through his first telescope. The four tiny "stars" changed their position from night to night, and Galileo became convinced that they were orbiting Jupiter. It was an extremely important discovery because at that time the heliocentric theory of Copernicus was still not widely accepted. Theorists and theologians were still battling to maintain the Earth in the center of creation. As long as the heliocentric theory depended solely on mathematical calculations and logical deductions, it was vulnerable. But Galileo provided irrefutable visual evidence that objects were revolving around a body other than the Earth. This was a devastating blow to those who believed for religious or philosophical reasons that the Earth was the center of everything. The Catholic Church recognized the threat and responded by placing Galileo's works on the Index of prohibited books. The church did not get around to reconsidering Galileo's "guilt" until the late 1970's—better late than never.

The moons of Jupiter were the first "new" astronomical objects ever discovered, but this was an age of terrestrial discovery as well, and Galileo followed the lead of the Balboas and Magellans. He claimed the right to name his discoveries. The politically astute scientist called his moons the Medicean stars, after the Grand Duke of Tuscany, Cosmo de'Medici, and his family. By analogy, it was as if a twentieth-century astronomer wanted to name his discoveries after the Rockefellers. (When William Herschel discovered the seventh planet, he wanted to name it after *his* sovereign; Uranus was almost named George.)

Galileo, however, had competition. The German mathematician Simon Marius independently discovered the Jovian satellites at about the same time, although it is questionable whether or not he realized

Unlike Earth with its single satellite, Jupiter is orbited by the four large moons and single small one shown, as well as by a dozen or more still smaller satellites.

what he was seeing. In any case, Marius also claimed the right to give names, and what Galileo called the Medicea Sidera, Marius called Sidera Brandenburgica. The whole matter might have degenerated into an Italo-German *Star Wars* but for the fact that non-Italian and non-German astronomers didn't like either set of names. Eventually, the four moons came to be known as the Galilean satellites. The individual names finally proposed by Marius, however, have survived. He suggested that the moons of Jupiter should be named after that god's mythological lovers, of which there were many. The four controversial moons were finally named Io, Europa, Ganymede, and Callisto.

In 1892, American astronomer E. E. Barnard discovered a fifth Jovian moon, much smaller than the others, inside the orbit of Io. It is unknown whether Barnard felt the temptation to name it after Grover Cleveland or Benjamin Harrison; in any event, the moon was given the less controversial name of Amalthea, after the she-goat who suckled the infant Jupiter. Between 1904 and 1974, eight other moons were discovered, all of them quite small and orbiting beyond the Galilean satellites. These were also named after Jupiter's Olympian lovers.

Even when seen through the largest modern telescopes, the Galilean satellites appear as little more than points of light. Before the 1970's, almost nothing was known about them, except that they were quite large, as moons go. Io and Europa are about the same size as Earth's moon, while Ganymede and Callisto are even larger, roughly the same size as the planet Mercury.

By carefully measuring the orbital motions of the Galilean satellites, it was possible to calculate their densities. Io and Europa both have densities comparable to that of our own moon (moon—3.34; Io—3.55; Europa—3.04, where the density of water equals 1.00), while Ganymede and Callisto are considerably less dense, weighing in at 1.93 and 1.84, respectively. None of the four, it was clear, could be another hydrogen-helium gas world. All four had to have solid, rocky surfaces, although the lighter density of Ganymede and Callisto implied the presence of a large amount of water in the form of ice. The differing densities seemed to be a clue to the early history of Jupiter. If the young Jupiter was a glowing, incandescent body, the heat it radiated would have "cooked" Io and Europa, driving out most of the light, volatile compounds that had not already escaped the slight pull of their gravity and leaving behind two dense, dessicated lumps of rock. Farther away from the hot proto-Jupiter, Ganymede and Callisto were able to retain their water, resulting in lower densities.

In the 1970's, data from sophisticated new instruments and techniques led scientists to suspect that the Galilean satellites might be more than just dead, inactive bodies like our own moon. Some of the Jovian radio emissions seemed to be tied to the orbital period of Io, a curious circumstance made even more remarkable by the discovery of a cloud of sodium in the vicinity of Io's orbit. Unlike Earth's moon, which spends nearly all the time outside Earth's magnetosphere, the four Galilean satellites orbit well within Jupiter's enormous and powerful magnetic field. The radio emissions and the sodium cloud suggested that the moons (at least Io) might be actively involved in some sort of electromagnetic interaction with Jupiter itself.

The moons of Saturn were even more interesting. Huygens discovered the first of them in 1655, a Mercury-sized satellite later named Titan. The name game was played at Saturn as well, and Cassini wanted to call the Saturnian moons the "Louisian Stars," after Louis XIV. That suggestion was not well received outside France, so for two centuries Saturn's moons were simply numbered rather than named. The numbering system was tied to the orbital distance of each satellite; but as new moons were discovered, some of the old moons had to be renumbered. The system was too confusing, so in 1858 Sir John Herschel took it upon himself to assign actual names, which are still in use.

Titan was long thought to be the largest moon in the solar system, but it is so far away that accurate measurements were difficult to achieve. Titan's main distinction, however, is that it is the only moon known to possess an appreciable atmosphere. In 1944, American astronomer Gerard Kuiper took a series of spectrograms that revealed the existence of a Titanian atmosphere composed of methane gas.

Far from the sun, Titan was likely to be very cold, with a surface temperature estimated to be around 125° Kelvin (−234°F). But the methane atmosphere was thought to be quite dense, perhaps as

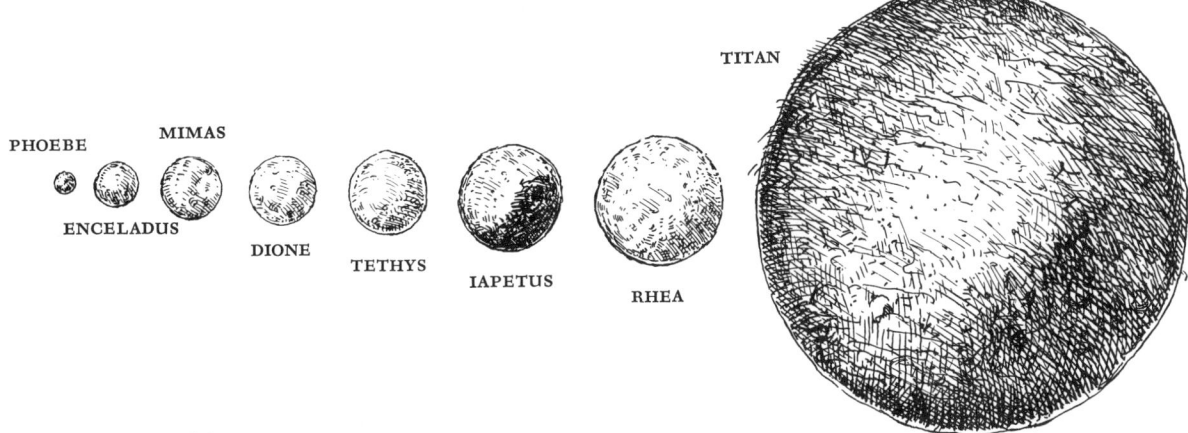

The Moons of Saturn

Titan appears gigantic next to Saturn's other satellites. Before Voyager, its actual size was not known, but it was thought to be the largest moon in the solar system.

dense as the Earth's. If that were the case, and if Titan had some internal heat source such as volcanism, then it was perhaps possible that the surface temperature could be much warmer than expected. Given methane, water, and a few warm spots on the surface, Titan might resemble the early Earth, where these same ingredients combined to form the first living organisms about four billion years ago. Some scientists theorized that Titan might very well be the best place in the solar system to look for life.

Most scientists considered this unlikely, although perhaps not impossible. Temperature was the key, and it seemed doubtful that Titan could be warm enough to support any sort of biological activity. The analogy to the early Earth, however, remained attractive. "Prior to 1980," writes atmospheric scientist James Pollack, "most planetologists concerned with Titan thought its atmosphere consisted largely of methane; they likened it to a primordial Earth, a terrestrial planet in a 'deep freeze.'"

Terrestrial analogies might apply to Titan, but for another Saturnian moon, Iapetus, there is no plausible counterpart anywhere else in the solar system. Cassini discovered this medium-size moon, about 1,600 kilometers in diameter, in 1671. A year later, he reported a curious fact about this new moon: it was visible only when it was west of the planet. When its orbit took it east of Saturn, it disappeared from view. Cassini concluded, "It seems that one part of his surface is not so capable of reflecting to us the light of the Sun which maketh it visible, as the other part is." In other words, part of Iapetus was dark and part of it was bright.

Later astronomers found that the trailing hemisphere of Iapetus measured six times brighter than its leading hemisphere. As astronomer William K. Hartmann describes it, one hemisphere is darker than a blackboard and the other hemisphere is as bright as snow. Although some other bodies in the solar system show variations in reflectivity from one hemisphere to another, nowhere else is the contrast so great.

Telescopic observations of Iapetus could not show enough detail to explain this strange situation. Many theories were proposed, but in the absence of close observations, the nature of Iapetus remained a mystery. One of the more plausible theories, proposed by Steven Soter of Cornell, suggested that the leading edge of Iapetus might be sweeping up very fine, dark debris eroded from the surface of a neighboring moon, Phoebe. A less plausible but undeniably more exciting idea about Iapetus was put forward by science-fiction writer Arthur C. Clarke. In the book version of 2001: A Space Odyssey, Clarke put his mysterious Stargate on Iapetus. During the Pioneer 11 encounter with Saturn, NASA arranged a phone link to Clarke, who was in London at the time. With reality about to catch up to his fiction, Clarke didn't try to defend his "theory." "I chose Iapetus for the Stargate," he said simply, "because it's a very peculiar body."

Peculiar, indeed. Out there, in the realm of giants, *everything* is peculiar. Our Earth-bred sensibilities are jolted by the wonders of Jupiter and Saturn. Seen from Earth through the unyielding veil of distance, there is much about Jupiter and Saturn and their retinue of moons that remains incomprehensible, unknowable. But by the beginning of the 1970's, scientists were getting ready to pierce that veil and for the first time see the giants as they really are. Scientists hoped that the close encounters of their spacecraft would answer their enduring questions about Jupiter and Saturn.

In that hope, they were not disappointed.

They were, however, surprised. Some of the answers would, in the end, turn out to be even more mysterious than the questions.

3 THE MISSION

The creed of aviation has always been simple: Faster, Higher, Farther. The driving force behind exploration of all types is equally simple: "I wonder what's on the other side of that hill?" If something so nebulous as the American "character" can be distilled to a single phrase, it would probably be, "Let's try something new..."

When the American space program was getting under way in the late fifties and early sixties, almost no one seriously spoke about Jupiter and Saturn. President Kennedy said we should go to the moon, but he said nothing about the rest of the solar system. Yet, in retrospect, it seems inevitable that once the bonds of Earth had been broken, we would be drawn ever onward by the challenges of distance, innovation, and curiosity. The voice of the Space Age was really the age-old song of the sirens.

Noel Hinners, director of the National Air and Space Museum,* describes the last two decades as "the Golden Age of Solar System Exploration." Historians have always found it a challenge to explain "golden ages," but Hinners lists "four dominant reasons" for the American planetary exploration program: national prestige, "vision," the search for knowledge, and "applications," that is, the economic spinoffs of space exploration. Ultimately, however, the imperatives perceived by historians must be expressed through the decisions made by individuals. Few people routinely think of themselves as instruments of history; they live their lives and do their jobs and let history take care of itself. The people who make history are seldom conscious of it.

The decision to explore Jupiter, Saturn, and the outer solar system was, in a sense, one of history's unconscious decisions. Historians will find plenty of documentation describing how the formal decision was arrived at and approved in 1969. Committee recommendations and Congressional appropriations are all in the record. What is not in the record, however, is the unconscious mental evolution of the previous dozen years that made the formal decision possible and probably inevitable.

The first American planetary missions were, in some respects, merely a response to Soviet missions. But it was clear that a successful exploratory program would have to be based on sounder principles than a reflexive "tit for tat." Something resembling a grand strategy for the exploration of the solar system had evolved by the mid-sixties, but it was highly vulnerable to the shifting winds of terrestrial politics. Manned space flight always had top priority at NASA, first with the moon program and later with the development of the shuttle. Planetary programs were not so easy to sell to Congress, even though they were far cheaper than manned programs, in part because they took longer to get results; a representative who voted to approve Voyager in the early seventies, for example, would have to wait nearly a decade to reap any political dividends. As Hinners points out, "Under ideal circumstances, the time between mission concept and launch is 5–10 years (in practice, more often 10–15 years). This flies directly in the face of the United States'

---

* In 1982, Hinners was named director of NASA's Goddard Space Flight Center in Greenbelt, Maryland.

yearly federal budgeting process and four-year Administrations. Add to this the competing considerations of relative science merit, technical feasibility, launch-vehicle availability, absolute and relative cost, and political and popular support—the result is a very inefficient process."

Thus, the NASA planetary strategy was really just a skeletal framework, a twenty-year wish-list that paid little attention to the frustrating realities of Planet Earth. The framework, at least, was logical. The first phase of solar system exploration was to consist of reconnaissance: simple, inexpensive missions designed to show us the planets and whatever else there was to be seen. The early Ranger, Pioneer, and Mariner missions were excellent examples of the reconnaissance phase. Next would come exploration, exemplified by the Viking landers on Mars, which returned detailed scientific data. The next level was to be intensive study, which would involve permanent manned bases in space or on the moon. The final hoped-for phase would be utilization, in which the moon and asteroids would be mined and various industries would move into space. Eventually, the planets might be colonized.

In the mid-sixties, as the reconnaissance phase was beginning, NASA began planning its second-level exploration missions for the seventies. In those palmy days when NASA got nearly everything it asked for from Congress, the planners thought big. The reconnaissance of the inner solar system was going well; we had seen the first fuzzy pictures of Mars and men would soon be standing on the moon. There seemed to be no reason for not extending the program to the outer solar system; and in an age of big boosters and big budgets, it seemed feasible to combine reconnaissance with exploration in a series of heavy-duty billion-dollar missions.

After Venus and Mars, Jupiter and Saturn beckoned. It was *possible* to go; therefore, we *would* go. What seemed fantastic at the beginning of the decade was virtually taken for granted by the end of it. Whatever the historical forces that prompted it, the original decision to explore space was open-ended, and the exploration of Jupiter and Saturn was simply a logical consequence of the decision to go to the moon. For the individuals who actually made the formal decisions, the real question about the exploration of the outer solar system became not *if*, but *how*.

The answer to the question of *how* seemed to be dictated by the planets themselves. The outer planets were approaching an alignment that occurs just once every 179 years. By the late seventies, Jupiter, Saturn, Uranus, Neptune, and Pluto would all be lined up on the same side of the sun. By taking advantage of the alignment, it was possible to design multiplanet missions that would encounter all five of the outer planets.

NASA called the concept the Outer Planets Grand Tour. The original plan envisioned twin launches to Jupiter, Saturn, and Pluto in 1976 and 1977, followed by twin launches to Jupiter, Uranus, and Neptune in 1979. The four-spacecraft, five-planet Grand Tour was to have cost about $750 million.

At the same time, NASA was planning a $2 billion Mars exploration mission known as Voyager. It was to consist of two Mars orbiters and two Mars landers, all launched aboard a mammoth Saturn V booster. In the frenzy of the race to the moon, all things seemed possible.

But the forces of politics, war, and social change combined to shoot down Voyager before it ever got off the ground. In a real sense, the space program was another casualty of the Vietnam War. The winds had shifted, and Congress was no longer inclined to give the space agency blank checks. Voyager was cancelled and replaced by the less ambitious Viking program.

The Grand Tour ran aground on the same reef. NASA wanted $30 million in startup funds for fiscal year 1972. Congress, which had already cancelled the last three Apollo missions, doled out just $10 million, and there were signs that money would be even harder to come by in the future. The NASA planners were forced to rethink their entire outer-solar-system strategy.

Uranus, Neptune, and Pluto got the ax. Because less was known about them than about Jupiter and Saturn, scientists had fewer specific goals in mind for the exploration of those worlds. More important, by eliminating the outer planets, several years were cut off the lifetime of the proposed mission. That meant that the spacecraft themselves didn't need to be so hardy, and that expensive ground support facilities wouldn't be necessary for an extended period.

Jupiter and Saturn remained. The restructured, scaled-down Grand Tour evolved into Mariner Jupiter-Saturn. Two Mariner-class spacecraft would be launched aboard Titan-Centaur rockets in 1977, targeted for Jupiter encounters in 1979 and Saturn encounters in 1980 and 1981. The total cost would be about $250 million, or about a third of the Grand Tour price tag.

In 1977, not without a touch of irony, NASA gave Mariner Jupiter-Saturn a new name—they called it Voyager, recycling the discarded name of the cancelled Mars mission and slapping it on an impoverished stepchild of the Grand Tour.

Planning a mission to Jupiter and Saturn is, in some respects, no different from planning a vacation trip to the Caribbean. The basic question is the same: *What should I take?* Knowing the destination resolves some questions (tennis shoes, yes; ski boots, no), but the prudent traveler must be conscious of weight restrictions, the possibility of losing a suitcase, the likelihood of bad weather, and the chance of sudden unplanned changes in itinerary. He should also have a clear idea of what he wants to do once he reaches his destination.

Explorers have an even more difficult task, since they have no tourist brochures to guide them. Some polar explorers brought horses with them, others took dog sled teams; and some men died because they made the wrong choice. Lewis and Clark, on the other hand, brought along just about everything they could think of, including a writing desk for Lewis. The heavily encumbered expedition moved slowly, but in three years in the wilderness, they lost but one man.

Deciding what to take to explore Jupiter and Saturn was a difficult matter. Although no lives were at stake, careers were, to say nothing of a quarter-billion-dollar investment. The Voyager spacecraft was to weigh 815 kilograms, but the scientific payload would weigh just 115 kg. It was not possible to follow Lewis and Clark's strategy of bringing along everything that could possibly be wanted. The instrumentation for the mission would have to be carefully chosen, small, efficient, reliable, and designed to provide a maximum return of scientific data.

There were four broad areas scientists wanted to study on a Jupiter-Saturn mission: the planets themselves, their magnetospheres, their satellites, and the environment of deep space (which scientists now prefer to call "the interplanetary medium"). A variety of instruments would be required to study these different objectives, and none of them could be off-the-shelf hardware. Each instrument had to function as part of the complete science package, within strict limits on weight, power use, and on-board computer capacity.

In 1972, NASA sent out an "Announcement of Flight Opportunity." Scientists from around the world responded with seventy-seven proposals for investigations, including thirty-one instrument designs and forty-six individual experiments. From these proposals, a NASA panel selected nine instruments to fly aboard Voyager. A tenth instrument, the imaging system, was assigned directly to the Jet Propulsion Laboratory for development. One additional experiment, radio science, required no specialized instrument other than the spacecraft's radio transmitter. Eleven principal investigators (PI's, in space science parlance) were selected to head the science teams, and the overall coordination of Voyager science was entrusted to Project Scientist Edward C. Stone, a magnetospheric physicist from the California Institute of Technology, which runs the Jet Propulsion Laboratory.

The Voyager science investigations fall into two general categories—those which rely on remote sensing and those which rely on direct sensing. The instruments for remote sensing were designed to collect data from light—in visible and nonvisible wavelengths—reflected or emitted from the planets and satellites. The direct-sensing instruments were built to study the immediate environment of the spacecraft itself. They were mainly concerned with high-energy radiation. Since the Voyagers would be flying through regions where they would be exposed to potentially damaging bombardment from high-energy subatomic particles, the direct sensors would provide, not just experimental data, but also diagnostic information about the health of the spacecraft.

In the view of the general public, the success of any planetary mission is judged by the quality of the pictures it returns. The early Mariner missions to Mars sent back only a few, rather grainy black-and-white pictures that, as it turned out, were scientifically misleading and not very interesting to the man in the street. By the time Viking reached Mars, imaging systems had improved dramatically and the pictures were exciting and spectacular to scientist and non-

THE MISSION

scientist alike. For Voyager, NASA and JPL designed the most sophisticated imaging system yet flown.

The Voyager imaging system consists of two television cameras mounted on the scan platform at the end of a steerable "science boom." Light from the target object passes through a series of lenses and mirrors and strikes a selenium-sulfur vidicon tube. The incoming light, in the form of photon particles, "excites" the atoms of the tube; the more energetic the photons, the greater the excitation. The energy levels across the one-inch tube are scanned and converted to electrical signals, with each signal representing the brightness level of individual picture elements, or pixels. A Voyager picture consists of 640,000 pixels arranged in an 800 by 800 square, and the signal from each pixel is converted to an eight-digit binary number. It is these numbers—not the picture—that are transmitted back to Earth. The computers at the Jet Propulsion Laboratory then reverse the process and translate the numbers back into 640,000 dots of the specified brightness. The end product is a sharp, clear picture. Because the Voyager images consist of nothing but numbers, they can be manipulated in a variety of ways to emphasize particular features.

Suppose a camera photographs (or, more correctly, "images") a dim, low-contrast object. Out of 256 possible brightness levels, this image consists mainly of pixels within a narrow range of, say, twenty brightness levels. The resulting picture would not reveal much detail. But the programmer can tell the computer to "stretch" the picture by giving new values to the brightness levels of each pixel. He can arbitrarily say that, for this image, brightness level 80 now equals brightness level 1 and brightness level 100 now equals level 256. The low contrast range of 20 numbers is thus expanded to 256 numbers. In the picture, dark pixels become darker, bright pixels become brighter, and the contrast is enhanced, bringing out details that might not otherwise be visible. For false color pictures, the procedure is much the same.

Each camera is equipped with a rotating filter wheel. The narrow-angle camera has six filters (clear, violet, blue, orange, green, and ultraviolet) and the wide-angle camera has eight, including three extremely narrow-band filters centered on important spectral wavelengths of sodium and methane. The variety of filters makes it possible to examine particular aspects of the planets and satellites in order to generate the maximum amount of useful data. For color pictures, three successive images of the same target are taken through blue, green, and orange filters. The digital information can then be manipulated by computer to produce either true or false color images.

The false color images, like the computer-"stretched" black-and-white pictures, are extremely useful to scientists, although they are sometimes misleading for laymen who see these processed images reproduced in newspapers or magazines. To the computer, the numbers that make up the pictures are meaningless. The computer programmer assigns meaning to the numbers, and the computer simply follows instructions. Depending on what meaning the programmer chooses to assign, the same set of numbers can be turned into radically different pictures.

The wide-angle camera has a 200mm focal length and 3-degree field of view. The narrow-angle camera has a 1500mm focal length and a field of view of 0.4 degree. Both cameras are necessary, since the disk of Jupiter would completely fill the narrow-angle frame some twenty days before closest approach to the planet. Planetary encounters, in fact, are divided into distinct phases dictated by the limitations of the imaging system. The observatory phase begins from long range and ends when the full disk of the planet can no longer be captured in a single narrow-angle frame. The far-encounter phase begins when 2 by 2 narrow-angle mosaics are necessary to capture the full disk. The near-encounter phase begins at an arbitrary moment, hours to days before closest approach to the planet; the Voyager 2 near-encounter phase at Saturn, for example, began sixteen hours before closest approach and ended twenty-eight hours after closest approach.

The principal investigator and leader of the Voyager imaging team is Bradford A. Smith of the University of Arizona. By the time of the Voyager 1 Jupiter encounter, twenty-two other scientists were members of the team. NASA prefers that its science teams speak with one voice, but the imaging team was more of a chorus—and not everyone was always singing the same tune. The team consisted of astronomers, geologists, and atmospheric scientists, each with his or her own ideas about what should be imaged and what the pictures meant. Although Smith was usually the team spokesman, other scientists contributed their own interpretations and analyses. In the end, though, the Voyager images really spoke for themselves.

Two of the primary goals of the Voyager mission were to determine the composition of the atmospheres of Jupiter, Saturn, and Titan, and to measure the internal heat flow of the two gas giants. The infrared spectrometer was designed to meet both those objectives.

Earth-based infrared astronomy, despite its inherent limitations, had identified a few of the constituent gases in the Jovian and Saturnian atmospheres. The Voyager *i*nfrared *i*nterferometer spectrometer (known as IRIS) would be able to find the signature spectra of other compounds, including the organic molecules some scientists hoped would be present. In addition, IRIS was designed to take data that would permit the mapping of heat flow within the planetary atmospheres. This was considered vital to any understanding of the belts and zones and overall structure of the gas giants' atmospheres. The versatile IRIS would also take infrared readings of the surfaces of the airless moons of Jupiter and Saturn. Different types and sizes of rocks radiate heat at different rates, and IRIS data would thus give an indication of what sort of rocks comprised the surface material of the satellites.

IRIS was designed and built by Team Leader Rudolph Hanel and his co-workers at the NASA Goddard Space Flight Center in Greenbelt, Maryland. Hanel is the grand old man of infrared spectroscopy, and IRIS was his masterwork. Unlike all previous space-borne infrared detectors, IRIS was not limited to a few preselected wavelengths. The Voyager instrument could take readings at more than 2,000 wavelengths, making it exceptionally valuable on a mission to planets as complex as the gas giants.

Like IRIS, the Voyager *u*ltraviolet *s*pectrometer (UVS) had as its main objective the determination of the composition and structure of the atmospheres of the target worlds. Working in the shorter end of the electromagnetic spectrum, where particles are more energetic, the UVS could observe spectra produced by chemical reactions in the upper atmosphere. Those reactions are caused by ultraviolet light from the sun interacting with upper-atmosphere molecules; the input of energy from colliding photons breaks apart some molecules and causes others to form. By taking inventory of these reaction products,

UVS would give scientists important data about the upper atmosphere.

The UVS instrument would also observe the interaction of the Jovian magnetosphere with the atmosphere and with the satellites. As electrically charged particles from the magnetosphere hit the upper atmosphere, they excite, or energize, the atmospheric molecules, which in turn produce ultraviolet auroras; analysis of the auroras can give information about the nature and energy level of the particles in the magnetosphere.

The UVS team leader is Lyle Broadfoot of the Kitt Peak National Observatory and the University of Southern California. By the time of the Jupiter encounters, he led an international team of fourteen scientists. Although their instrument was designed to observe the planets, it was, as a bonus, useful for stellar astronomy as well. During the long months of cruising between worlds, the UVS carried out successful observations of ultraviolet sources far beyond the solar system.

Although scientists had been observing radio emissions from Jupiter since the mid-fifties, Earth-based receivers were generally unable to pinpoint the exact source of those emissions. Some of the short (centimeter) wavelength radio emissions were known to be thermal in origin, a product of Jupiter's primordial heat reservoir. But the strong radio pulses at decimeter wavelengths and the unpredictable, highly energetic emissions at meter and decameter wavelengths were still major mysteries.

The *p*lanetary *r*adio *a*stronomy (PRA) experiment on Voyager was designed to clear up some of those questions. The instrument consists of two 10-meter antennas, extending from the spacecraft at right angles to each other, and a receiver that can cover a broad range of frequencies from 1.2 kiloHertz to 40.5 megaHertz. The PRA team leader is James W. Warwick, an astronomer from the University of Colorado at Boulder; eleven other scientists from the U.S. and France complete the team.

Warwick had been studying Jovian radio emissions since 1960, and he had some definite ideas about what Voyager would be likely to find. He was especially interested in planetary magnetic fields, and had worked out a deductive theory to predict what fields are like on the outer planets. Warwick's planetary radio astronomy experiment would give important data on the strength of the magnetic fields at Jupiter and—presumably—Saturn. Some weak radio emissions from Saturn had been observed in 1975 and implied the existence of a rather small Saturnian magnetic field, but the magnetic and radio properties of Saturn were still almost completely unknown.

The PRA team also hoped to be able to detect lightning on Jupiter and Saturn. Theory predicted that lightning should be present in their atmospheres, and possibly on Titan as well. While the imaging team looked at the night side of the planets for visual evidence of lightning, the PRA would listen for the characteristic "snap" of electrical discharge, similar to the radio static one hears during a thunderstorm.

The final remote-sensing instrument aboard the Voyagers was the photopolarimeter. By analyzing the way in which light is reflected from the Jovian and Saturnian clouds, the photopolarimeter would give important information on the size, distribution, and nature of the particles present in the upper atmospheres. It could also contribute data about the particles in Saturn's rings.

Unfortunately, the Voyager photopolarimeters did none of those things for most of the journey. The instruments were relatively simple in design, but Murphy's Law applies in space as well as on Earth. On Voyager 1, two filter wheels stuck on the photopolarimeter; later, the electronics broke down, too, and the instrument was turned off. Voyager 2 experienced similar problems, and the operation of the instrument was severely restricted.

It is difficult to think of a planetary mission in which one or more of the science instruments did not break down. "Glitches" are simply a part of the business. An instrument that works perfectly in the laboratory may refuse to function at all in the demanding environment of outer space. Despite multiple redundancy in design, rigorous testing, and even a few successful long-range "fixes," when an instrument aboard a spacecraft breaks down in flight, there is very little that can be done about failures. For scientists who have spent years developing instruments and planning experiments, such failures are maddening. During planetary encounters, the scientists whose instruments are still healthy speak of the failed experiment as one

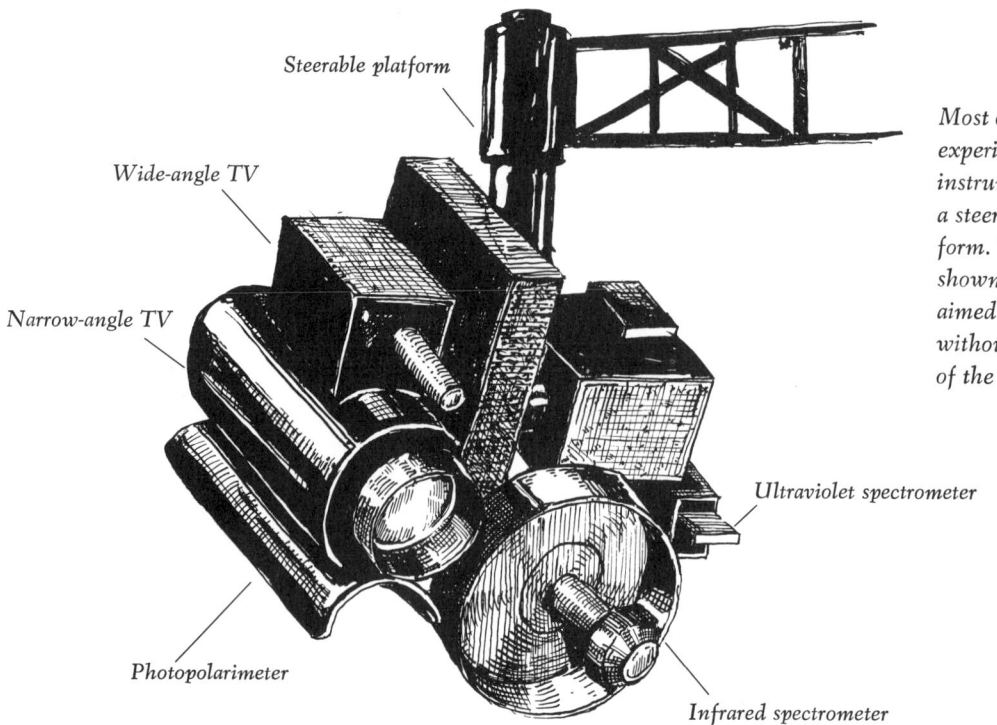

Most of Voyager's major experiments depend on instruments mounted on a steerable scan platform. The instruments shown here can be aimed at their target without moving the rest of the spacecraft.

Voyager's Major Instruments

might speak of a family member who has contracted an incurable disease; they regret the loss and sympathize with the victim, but down deep, one suspects, they are simply glad it didn't happen to them.

Voyager's remote-sensing instruments look at the electromagnetic spectrum in a variety of wavelengths. But those same waves can also be interpreted as collections of subatomic particles; in modern quantum physics, waves and particles are simply different aspects of the same phenomenon. Voyager's direct-sensing instruments are designed to analyze the particles and the electrical and magnetic fields through which they travel.

When a particle collides with one of Voyager's direct-sensing instruments, it transfers energy to the detectors. The amount of energy transferred tells scientists the nature of the particle and allows them to deduce its probable origin. Subatomic particles emanating from the sun, for example, have a distinctive electrical charge and energy level which differ from those of particles originating within a planet's magnetosphere.

For physicists, Jupiter is the Land of Oz. Jupiter's immense magnetosphere, powerful magnetic field, and energetic interactions

THE MISSION

with Io make it a dream world for anyone interested in the behavior of particles and fields. The origin and nature of the sodium cloud near Io—known as a torus after the mathematical term for a doughnut shape—were not well understood before Voyager. Another oddity to be explained was an apparently unique electromagnetic phenomenon known as the Io flux tube. Magnetic lines of force emerging from Jupiter pass through Io and connect the two bodies with a powerful electric current. The strength of the current was estimated to be about a million amperes, and Voyager 1 was aimed to fly right through this powerful flux tube.

Jupiter also interacts strongly with the solar wind. The sun continuously shoots out streams of a wide variety of subatomic particles, collectively known as the solar wind, which literally fill the solar system. The existence of these particles is one reason why scientists now prefer to call space the "interplanetary medium"; "space" implies emptiness, and the solar system is anything but empty. When the solar particles collide with the Jovian magnetosphere, they ricochet from the planet, but not before their impact distorts the shape of the magnetosphere. The same thing happens with the Earth's much smaller magnetosphere. Especially powerful particles may overwhelm

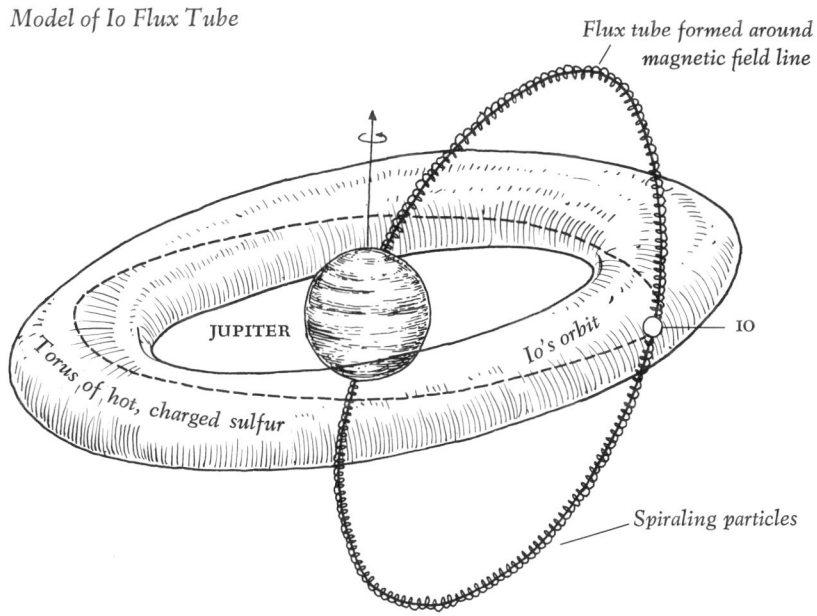

Model of Io Flux Tube

Jupiter is encircled by a torus, or doughnut-shaped cloud of hot, electrically charged sulfur particles. As Io moves through the cloud, it generates an intense electrical current known as the Io flux tube, connecting Io and Jupiter.

Earth's magnetic field and spiral inward toward the planet along the field lines, producing spectacular auroras in the upper atmosphere. Scientists suspected that Jupiter might have auroral displays of its own, in the visible spectrum as well as in the ultraviolet.

Five direct-sensing instruments were placed aboard the Voyagers to collect data on the particles and fields in the neighborhood of Jupiter and Saturn, and in the depths of the interplanetary medium. They included a magnetometer experiment, two plasma experiments, an instrument designed to measure electrically charged particles, and a cosmic ray detector.

The magnetometers (there are three) are affixed to a long 13-meter boom extending from the spacecraft. The boom is necessary to isolate the instruments from the magnetic fields created by Voyager's own electronics. The magnetometers could not measure the strength of the magnetic fields *on* Jupiter (or Saturn, or their satellites); rather, as direct-sensing devices, they measured the field in the immediate vicinity of the spacecraft. However, the raw data collected could be analyzed in such a way as to give an indirect measurement of the intrinsic strength of the planetary fields. Norman F. Ness of NASA Goddard is the team leader.

Of the two plasma experiments, one looks for plasma particles and the other looks for plasma waves. Plasma is the fourth state of matter (liquid, solid, and neutral gas are the others), and it consists of a swarm of electrons and protons. At very high temperatures, atoms are stripped of their negatively charged electrons, leaving the positively charged protons in the nucleus. However, in a given volume of plasma, the total numbers of electrons and protons are equal, giving the plasma a net electrical charge of zero.

The plasma-particle experiment is led by Herbert S. Bridge of MIT. The experiment was designed to collect data on the population, density, and temperature of the plasma particles in the Jovian magnetosphere. En route, the instrument could also study the solar wind. Unlike the remote-sensing instruments, the plasma instrument could not be aimed. In order to change its orientation, the entire spacecraft had to be maneuvered.

When energy (such as from solar wind particles) hits plasma, waves or oscillations may be generated. Analysis of the waves may yield further data on the density and temperature of the plasma. The

Voyager plasma-wave experiment gets its data via the same two metal rods employed as antennas by the planetary radio astronomy experiment. Leader of the plasma-wave team is Frederick L. Scarf, from TRW, Inc. Scarf is the only PI on Voyager to come from private industry—all the others are associated with government or universities.

The *low-energy charged particle* (LECP) experiment actually measures particles whose energy is *higher* than can be measured by either of the plasma experiments. However, in the world of particle physics, the particles detected by the LECP, traveling at a few percent of the speed of light, are considered to be lethargic slowpokes.

The LECP uses two detectors "tuned" to different energy ranges. One measures particles trapped in the magnetosphere having energies of up to 160 million electron volts (MeV). The other detector is a kind of telescope for remote sensing in areas of low particle density. It covers a range from 0.15 MeV to about 10 MeV. The LECP team leader is Stamatios (Tom) Krimigis, who comes from Greece by way of the Applied Physics Laboratory at Johns Hopkins University.

The particles from the sun and the particles in planetary fields are not the only ones Voyager would encounter. The entire solar system is bombarded by very-high-energy charged particles that are born in mystery somewhere out in the depths of space—in this case, the inter*stellar* medium. The solar wind "shields" the solar system from these particles in much the same way that a magnetosphere shields a planet from solar particles. These interstellar particles are known as cosmic rays, although the "rays" (or waves) are produced only when the particles collide with something, such as air molecules in our upper atmosphere.

The Voyager cosmic ray experiment, led by Rochus E. Vogt of Caltech, is the only instrument on board that is not really concerned with the planets. Although the experiment would observe interactions between cosmic rays and the planetary magnetospheres, its real purpose is to measure the flow of cosmic rays beyond the orbit of Saturn, where the sun's influence declines. The instrument can measure and analyze particles with energies of up to 500 MeV; a particle with such a high energy level might be traveling at ninety-nine percent of the speed of light.

The final Voyager experiment isn't really an experiment—it just *happens*. To keep in touch with Earth, a spacecraft broadcasts on radio frequencies. If the spacecraft goes behind a planet, as seen from Earth, its signal is cut off, or occulted. But if the planet has an atmosphere, the occultation is gradual. By analyzing the way in which a radio signal is occulted, scientists can deduce a great deal about the nature and structure of the atmosphere. The propagation of radio waves is slowed down by gas molecules which are electrically neutral and speeded up by electrically charged particles in the ionosphere, the upper atmosphere layer where charged particles congregate. The density of the gas or charged particles determines the degree to which the radio waves are affected. The radio occultation data was expected to yield temperature–pressure profiles for the upper atmospheres of Jupiter, Saturn, and Titan. The Voyager trajectories were designed to assure that the occultations would take place. The team leader for radio science is Von R. Eshleman from the Center for Radio Astronomy at Stanford.

These eleven experiments and the instruments that support them, it might be noted, are generally biased toward Jupiter. Since much less was known about Saturn, it would have been more difficult to design experiments to answer specific questions about it. Jupiter was much more intensively studied, so there was less uncertainty about what to take along on a Jupiter expedition. There was also the unpleasant possibility that the Voyagers would not reach Saturn in operating condition; it was conceivable that the high radiation at Jupiter might knock out most of the instruments. In effect, NASA hedged its bets by loading the Voyagers with Jupiter-intensive experiments. If the spacecraft did make it to Saturn, so much the better; instruments designed for one gas giant ought to be adequate for a preliminary study of another gas giant.

The science instruments aboard the Voyagers were only as good as the spacecraft that carried them. As the original name of the mission (Mariner Jupiter-Saturn) suggests, the Voyagers are descendants of the Mariner-class spacecraft that explored the inner solar system so effectively in the sixties and early seventies. The Voyagers more than lived up to their family tradition.

The guts of the spacecraft are contained in a ten-sided aluminum framework mounted with ten electronics compartments.

Ungainly as it looks, the Voyager spacecraft is actually a model of efficient design. Weighing just 815 kg at launch, the spacecraft carries a 115-kg science payload. With the extendable science and magnetometer booms in stowed position for launch, the spacecraft would fit comfortably in the average living room.

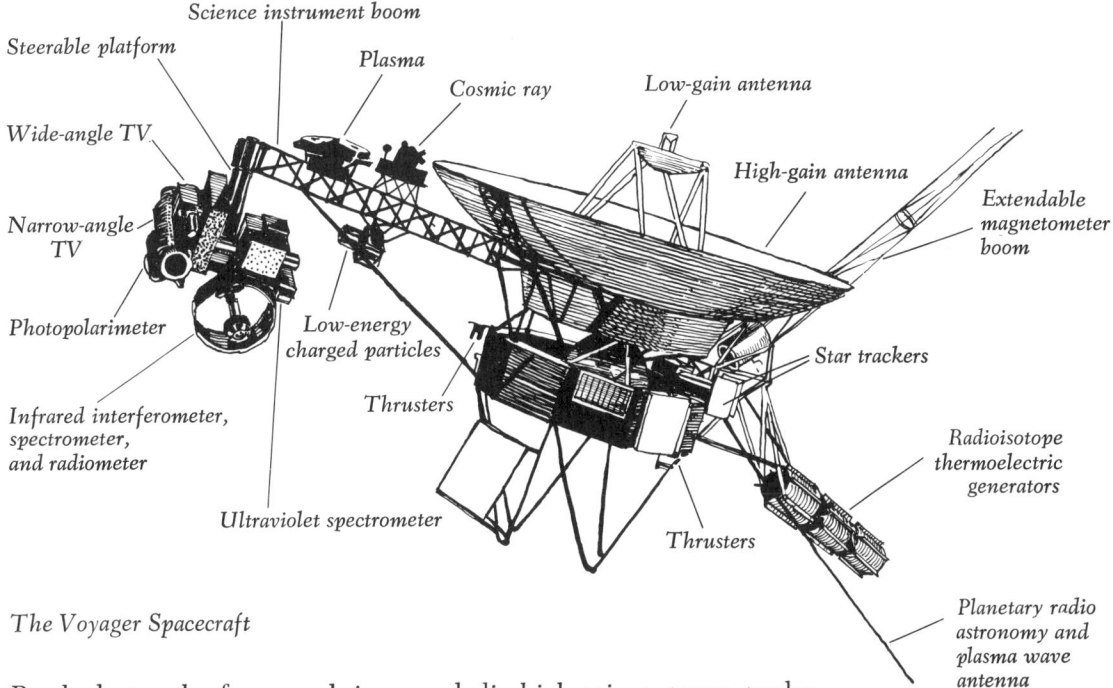

The Voyager Spacecraft

Perched atop the framework is a parabolic high-gain antenna twelve feet in diameter that is normally aimed at the Earth during flight. Five booms and antennas project outward from the frame at odd angles, giving the entire spacecraft the look of a hobby kit bristling with spare parts and accessories included as an afterthought.

Voyager doesn't look like its Mariner ancestors that explored the inner solar system because it has no solar panels. Close to the sun, solar energy is plentiful and easy to collect. The Mariners that went to Venus and Mercury needed only two solar panels, while the Mars Mariners had four. But the Voyagers, bound for regions where the light of the sun is less than one percent as intense as it is at Earth, could not rely on the sun. Instead, they use a form of nulear energy.

The power source that provides electrical energy to run Voyager's instruments and computers is an array of three *radioisotope thermoelectric generators* (RTG) mounted in canisters at the end of a steel and titanium boom. Energy comes from the natural radioactive decay of plutonium-238. The heat released by the plutonium is converted to electrical energy. At launch, the nuclear generators produce about 450 watts, but by the time of the Saturn encounter, some four years later, some of the plutonium has been used up and power output is down to about 400 watts. That is only enough to supply about a half-dozen light bulbs, but it is more than enough to keep the spacecraft ticking. The Voyager electronic subsystems are models of energy conservation; the ultraviolet spectrometer, for example, uses just 2.5 watts of power. Even the imaging system requires only 41.9 watts, or less than most light bulbs.

More than any previous spacecraft, the Voyagers had to be able to run themselves. Because of the great distance to Jupiter and Saturn, "real-time" control from Earth is a physical impossibility due to the cosmic limit of the speed of light. The one-way communication time to Jupiter is about forty minutes, and to Saturn, twice that. The spacecraft had to have versatile and independent computer brains that could operate with or without instructions from Earth.

The Voyager computer command subsystem consists of two independent and redundant 4,096-word memories. About half the computer capacity is used to store fixed "housekeeping" routines for normal operations. The remaining capacity is used for "uplinks" from Earth, messages sent "up" to the spacecraft with instructions for trajectory changes, engineering routines, and science observations. A single 1,290-word uplink sequence can generate as many as 300,000 individual onboard computer commands.

The Voyager computer, like the human brain, consists of two "lobes." Each is capable of checking the other for malfunctions; if either lobe fails, the other lobe automatically turns it off before it can issue erroneous commands to the spacecraft.

The computer can record, store, and transmit a truly awesome amount of information. Mariner 4, during its 1965 flyby of Mars, could transmit just $8\frac{1}{3}$ bits of data per second (a bit or "binary digit" is the basic unit of computer-talk). The Viking Orbiters, eleven years later, could send 16,000 bits per second (or 16 kilobits). The Voyagers, beneficiaries of swiftly evolving computer technology, could transmit 115,200 bits per second from Jupiter and 44,600 bps from Saturn. This immense capacity was necessary, considering that a single Voyager image consists of 5,120,000 bits, or more than were collected from the entire Mariner 4 mission. The onboard computer can store up to 100 5-million-bit pictures. The other science instruments require less computer capacity, although at peak performance the planetary radio astronomy experiment can eat up nearly as many bits as the imaging system.

All those bits are beamed back to Earth, where they are detected by the antennas of the Deep Space Network. With stations in Goldstone (California), Madrid (Spain), and Canberra (Australia), the DSN can keep a continuous watch on the Voyagers and other deep space probes. Once the spacecraft's signals have been received, they are amplified and relayed to the Space Flight Operations Facility at the Jet Propulsion Laboratory. There, the JPL computers process the data and print it out for the scientists and engineers.

The most complex aspect of the Voyager mission was the selection of trajectories for the two spacecraft. Getting the maximum science return required delivering the spacecraft to preselected points in space 500 million to a billion miles away, with arrival times precise to within minutes, two to four years after launch.

The original Grand Tour concept had been made possible by a clever navigational maneuver known as the gravity-assist technique. As a spacecraft approaches a planet, the gravitational field of the planet causes it to pick up speed; the closer the spacecraft comes to the planet, the more speed it gains. The spacecraft actually steals a bit of the planet's gravitational energy. For the spacecraft, the effect is dramatic; for the planet, the effect is somewhat more subtle. The two Voyager encounters actually slowed down the motion of Jupiter —by about one foot per trillion years. Knowing the mass of the planet and the speed of the spacecraft, it is possible to select trajectories that will use the gravitational speedup to fling the ship past the first planet and on toward another world. Without the gravity assist, missions to the outer planets would be virtually impossible with present-day technology.

The Voyager mission design called for gravity assists at Jupiter to get to Saturn. At Saturn, another option presented itself. Depending on timing and trajectory, Voyager 2 could repeat the gravity-assist technique at Saturn and continue on to Uranus, arriving there four and a half years later. And at Uranus, yet another gravity-assist option was available for a three-year trip to Neptune. Looking at these trajectories, the NASA planners realized that, with a little luck, they might get their Grand Tour after all.

However, the Voyager spacecraft were not the Grand Tour spacecraft. They were not designed to survive eight and a half years in space for a Uranus encounter, or the twelve years needed for a Neptune flyby. Still, the history of most planetary probes was that they usually greatly exceeded their expected life-spans. If Voyager 2 reached Saturn in good working order, the Uranus option might be feasible. Keeping the Uranus option alive, however, made the choice of flight paths that much more complicated.

The mission designers started with a list of science objectives for the Voyagers, then worked to find trajectories that would meet those goals. At Jupiter, they wanted to penetrate the magnetosphere, fly through the Io flux tube, get close enough to the planet for good imaging of the atmosphere and, particularly, the Great Red Spot, and get close to medium-range flybys of all four Galilean satellites. At Saturn, the goals included flying behind the rings, encounters with several satellites, and an extremely close pass at Titan.

In order to meet the science objectives, mission controllers would have to keep a close eye on fuel requirements. Once launched by the Titan Centaur booster, the Voyagers would have only a limited capability for altering their courses. The Voyager spacecraft has sixteen thrusters, four of which are used for trajectory changes, and all of which are supplied from a single tank containing 230 pounds of hydrazine fuel. After allowing for adjustments in the attitude of the spacecraft, science maneuvers at the planets, and normal midcourse corrections (the pressure of sunlight photons would cause the spacecraft to drift), there was not a very large fuel margin.

Voyager Mission Design Manager Charles Kohlhase, Navigation Team Leader Ed McKinley, and their co-workers spent about three years sorting through the possible trajectories to come up with the two that would yield the greatest science return. "We checked 300 to 400 trajectories in two or three years," said Kohlhase, "and made a lot of trade-offs." The designers picked the best possible Titan encounters, then worked their way back to Jupiter. They decided it would be best to encounter Titan on the way in, before the closest pass at Saturn; that way, the Titan data would be acquired even if the spacecraft should collide with the rings. After choosing the optimum Titan encounters, 50 to 100 trajectories remained.

The Jupiter encounters also entailed risk to the spacecraft from radiation. Voyager 1 would pass much closer to Jupiter than would Voyager 2, so it was in greater jeopardy. It was estimated that Voyager 1 ran a thirty percent chance of suffering a significant failure due to the Jovian radiation; for Voyager 2, the risk was estimated at only four percent. The high hazard required flexibility. If one space-

THE MISSION

craft should be lost, the other would have to be retargeted to make up for the missed science opportunities.

Kohlhase and his team picked through the attractive trajectories for Titan and the Galilean satellites, and found two that seemed especially good. As "sort of a fallout," said Kohlhase, they found that in the best trajectories, Voyager 1 would look at one side of each of the four Galilean satellites, and Voyager 2 would see the opposite side. Since no one had ever seen *either* side of these moons, seeing *both* sides would be a fantastic stroke of good luck; in some strange way, the solar system seemed to be cooperating with NASA.

To appreciate the complexity of choosing trajectories that would satisfy all the requirements, it helps to step back and take a broad view of the solar system. Earth, the Voyager launch platform, is hurtling around the sun at a velocity of 29.79 kilometers per second. Jupiter, half a billion miles away, is moving at 13.06 km/sec, and Saturn, a billion miles away, has an orbital velocity of 9.64 km/sec. The Galilean satellites orbit Jupiter with periods of between 2 and 16 days at distances ranging from 412,600 km (Io) to

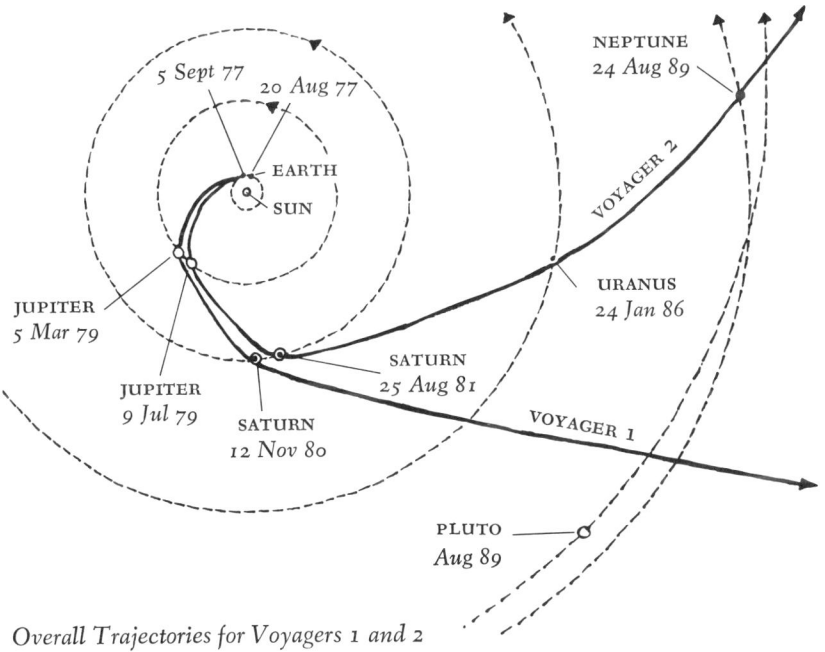

Overall Trajectories for Voyagers 1 and 2

The Voyagers' route to the stars was made possible by a rare alignment of planets which occurs only once every 179 years. Gravity assists at each planet help swing the spacecraft on to the next.

Last of its species, the multiengined Titan Centaur carried each of the Voyagers into space. Future probes will be launched aboard the reusable shuttle. The spacecraft, with booms retracted, fits neatly in the topstage nosecone.

Launch of Voyager 2, 20 August 1977

1,880,000 km (Callisto). Titan orbits Saturn at a distance of 1,222,000 km, with a period of just under 16 days. Everything is constantly in motion. Add to this consideration the fact that there was a basic uncertainty about the precise position of all these objects, simply because they are so far away from Earth. For the Galilean satellites, the uncertainty was about 300 to 400 km. Jupiter itself turned out to be some 200 km farther south of the plane of the ecliptic than was thought. These uncertainties seem small, but so was the margin for error. Hitting the aiming point for the Uranus option, for example, required an accuracy of plus or minus 2,800 km at Saturn.

Long-range accuracy depended on the onboard optical navigation system. The Voyagers were equipped with star trackers, generically known as Canopus trackers, after the bright star most spacecraft use as a navigational reference point. For complete accuracy, the star trackers needed to see stars as dim as ninth magnitude.* In practice, they registered stars as faint as magnitude 9.8. "If the optical navigation system works," said Kohlhase before the Voyager 1 Jupiter encounter, "getting to Uranus will be duck soup."

Before the Voyager Jupiter encounters, NASA soft-pedaled all talk of the Uranus and Neptune options. Kohlhase was about the only person who wasn't reluctant to mention the outer planets. "All of our thinking is still success oriented," he maintained. The key to Uranus was Titan. If Voyager 1 had a good encounter and satisfied the science objectives, then Voyager 2 would be free to attempt the Uranus option. If Voyager 1 failed at Titan, Voyager 2 would have to be retargeted from the Uranus aiming point for a closer Titan encounter, and the Uranus option would disappear. Kohlhase, two years before the Titan encounters, wasn't worried; he was already talking about the aiming point at Uranus for the Neptune option.

But Neptune, two billion miles from the sun, will not be the end for the Voyager mission. The spacecraft will continue on, toward interstellar space. At a distance of 50 to 100 AU (*astronomical units*: 1 AU is equal to the distance from the Earth to the sun), they will pass the heliopause, the outer boundary of the sun's influence. At that point, the solar wind is overwhelmed by the "galactic wind" of a hundred billion stars. With any luck, the spacecraft will still be sending back data when the boundary is crossed.

Eventually, the generators will use up their plutonium and, power gone, the Voyagers will fall silent. But their incredible journey will continue. Covering a distance of about 3 AU per year, the spacecraft will slowly move out into the Milky Way galaxy of which our sun is but one of billions of stars and from which neither our planet nor our spacecraft can ever escape. Voyager 1 is destined to pass within 1.7 light-years of a star in the constellation Ursa Minor, some

---

* Astronomers classify stars by brightness, or magnitude. The lower the magnitude, the brighter the star. The naked-eye limit is about sixth magnitude, which is roughly forty times brighter than the ninth magnitude stars seen by Voyager.

40,341 years after launch. Voyager 2 will pass within 1.1 light-years of the same star 44,426 years after launch. There will be other encounters even further in the future; Kohlhase and Paul Penzo have plotted stellar flybys for the next two million years.

Two million years, though, is just the blink of an eye in the history of the universe. The Voyagers will wander throughout our galaxy for *billions* of years—forever, in fact.

There is a chance—exceedingly small, but finite—that the journey of the Voyagers will be interrupted. Somewhere out in the depths of the galaxy, there may be other curious, clever, space-faring races. It is not inconceivable that on some distant day, one of those races will encounter a drifting derelict. They may take the ancient Voyager aboard their own spaceship for a closer examination. They will have many questions. They will want to know where the derelict came from, who launched it, why, and when. They will want to know what sort of beings we were.

And Voyager will tell them.

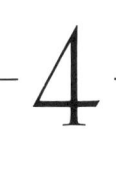 THE MESSAGE

In the glittering expanse of our Milky Way galaxy there are more than a hundred billion stars—perhaps a half dozen stars for every human being who has ever lived. Beyond the Milky War, scattered from here to the end of the universe, are billions of other galaxies whirling through time and space like the joyous glow of an endless celebration of existence. Yet, in all of it, in a universe bursting with matter and energy, we know of only one time and place where the elements have arranged themselves in such a way as to produce entities capable of understanding and appreciating the glory of existence. The place is Earth; the time is now.

It seems incredible that we should be the only sentient beings in such a vast and ancient universe. If one believes in God, then it must have have been an extravagant and wasteful God to have created so much for the contemplation of so few. And if one thinks of the universe as an immense physical laboratory, it is difficult to believe that an experiment that worked so well here was not tried elsewhere. There must be Others out there, somewhere....

If intelligent life does exist elsewhere in the universe, we have yet to discover it—or they, us. The distances between stars are so immense that an accidental encounter between species is wildly improbable. Contact will require a systematic effort—by them or by us, or by both.

Terrestrial attempts at interstellar contact, however, have been sporadic and ineffectual. Aside from a few deluded or opportunistic individuals who claim to have been whisked away aboard flying saucers—frequently crewed by sexy blonds or androgynous gods—no one on Earth has ever spoken with a being from another world. We don't *know* that they exist; we can only hope.

That hope gave rise to the Voyager interstellar record. Attached to each of the Voyager spacecraft are two gold-plated phonograph records that contain a kind of self-portrait, in words, pictures, and sounds, of the planet Earth. The record is really no more than a note in a bottle, and the chance that it will one day wash ashore on an inhabited stretch of beach in the cosmic ocean is, to be charitable, remote. But it *could* happen.

If it does happen, the people who put the note in the bottle will never know about it. They, like the rest of us, and perhaps the Earth itself, will be long dead. So the Voyager record is also a sort of time capsule, a carefully packaged archeological relic for unknown alien scholars to ponder in some unimaginable future.

Notes in bottles and time capsules are not typical NASA projects. The space agency is, after all, a bureaucracy, and bureaucracies are not noted for making grand symbolic gestures. But this gesture seemed worth the effort, for many reasons. The record may have been aimed for the stars, but the only certain recipients of the Voyager message are the people of Earth. For NASA, reaching earthlings was even more important than contacting extraterrestrials. The record was a way to focus attention on Voyager and NASA and to remind people of our most fundamental reasons for exploring space in the first place.

The Voyager record was not NASA's first attempt at interstellar (or intraterrestrial) communication. Pioneers 10 and 11, the first probes sent to Jupiter and Saturn, and launched in 1972 and 1973, each bore a simple plaque designed to show extraterrestrials who had launched the spacecraft, where, and when. The two Pioneers were to be the first manmade objects to leave the solar system, and it seemed appropriate that they should carry some sort of galactic ID tag.

Pioneer 10 encountered Jupiter in December of 1973, returned valuable data, then sailed onward toward the edge of the solar system. Pioneer 11 flew past Jupiter in December, 1974, then utilized the gravity-assist technique to swing almost completely around and backtrack across the inner solar system to get to Saturn, on the opposite side of the sun at the time, in September, 1979. Following the Saturn encounter, Pioneer 11 also continued on toward interstellar space. Both spacecraft will leave the solar system sometime in the 1990's.

The Pioneer plaques were the result of a collaboration between two Cornell astronomers, Frank Drake and Carl Sagan. Drake, a radio astronomer and director of the Arecibo Observatory (the world's largest radio telescope), is a long-time advocate of communication with extraterrestrial intelligence, or CETI. In the early 1960's, he led the first systematic effort to listen for radio signals from other star systems, Project Ozma. Later, in 1974, he was responsible for the first signal intentionally sent to other stars, a brief coded message beamed outward from Arecibo.

At the time the Pioneer plaques were created, Carl Sagan was not yet a best-selling author and television personality. He was primarily a planetary astronomer, known for his work on the atmosphere of Venus and the surface of Mars. He did, however, have a reputation as a scientist who was willing to speak up on controversial issues, such as UFO's and the apocalyptic theories of Immanuel Velikovsky. But Sagan's primary interest was (and is) the search for extraterrestrial life.

Both Sagan and Drake had spent a lot of time thinking about extraterrestrial life—what it might be like, how common it might be in the galaxy, and how we might communicate with it. Drake had even worked out an equation which sought to estimate the number of intelligent, technological civilizations in the Milky Way: $N = N_* f_p n_e f_l f_i f_c f_L$. The factors in the equation include the number of stars in the galaxy, the fraction of stars with planets, the fraction of planets with a biologically suitable environment, the fraction of those planets on which life actually arises, the fraction of those worlds on which life evolves into intelligence, the fraction of planets on which an intelligent species creates a technological civilization, and the fraction of a planet's lifetime during which a technological civilization exists—all multiplied by each other.

Drake's equation combines one reasonably well established number (the number of stars in the galaxy) with a few logical deductions and several blind guesses. The crucial number in the equation turned out to be the final one—the lifetime of technological civilizations. We have only one example of such a civilization, and there is a very real possibility that it will destroy itself after only a few decades of existence. If that is a typical circumstance, then the number of technological civilizations in the galaxy at any given moment must be very low, perhaps between one and ten. But if only one percent of all civilizations manage to survive their adolescence and go on to flourish for thousands or millions of years, then N might be a very large number; there could be millions of civilizations out there, waiting for us to join them.

An equation that produces an answer with an uncertainty of several hundred million must be viewed with a certain amount of skepticism. Princeton theoretical physicist (and CETI advocate) Freeman Dyson has written, "I reject as worthless all attempts to calculate from theoretical principles the frequency of occurrence of intelligent life forms in the universe." Dyson argues that multiplying one unknown quantity by another doesn't really tell you very much. Yet Dyson, like Drake and Sagan, is a firm believer in the existence of extraterrestrial intelligence—in unknown numbers at unknown locations.

Even if the galaxy is teeming with space-faring species, it is extremely unlikely that either the Pioneers or the Voyagers will ever be found by anyone or anything. Nevertheless, Sagan and Drake worked seriously to create a message that could be understood by anyone who might someday stumble across the path of the Pioneers. The plaques, made of gold-anodized aluminum, six by nine inches, contain a galactic "map" showing when and from where the spacecraft was launched, a diagram of the solar system (inaccurate, as it turned out), and a drawing of two human figures, male and female.

When the Pioneer plaques were made public, they aroused a

bizarre controversy that centered on the drawings of the male and female figures. The drawings depicted two vaguely pan-racial individuals, with the man raising his right arm to wave hello. Both figures are nude and anatomically detailed.

Blue noses from every continent immediately protested that NASA was sending "smut to the stars." Any decent extraterrestrial, it was argued, would be shocked by the sight of these wanton, naked Earthlings; we were inviting a raid by the Intergalactic Vice Squad. How could NASA squander taxpayers' dollars on X-rated spacecraft?

There were other lines of protest. Feminists complained that the woman in the drawing was smaller than the male and seemed to be standing slightly behind him—and why did the *man* get to wave hello? The pan-racial aspects of the drawing offended some people. The man had a broad, Negroid nose and wavy Caucasian-style hair, while the woman's eyes looked characteristically oriental. This was all intentional, but some saw it as an odious mongrelizing of the human race.

A respected British astronomer even argued that NASA had made a catastrophic mistake by revealing our location to aliens who might want to come down to Earth and enslave us, exterminate us, or simply eat us. This protest ignored the fact that we have already announced our existence to the galaxy via our television broadcasts. An expanding shell of TV signals, now about seventy light-years in diameter, is even now carrying *I Love Lucy* and *The Dukes of Hazzard* to the stars. If anyone out there really wants to invade Earth, it will probably not be due to the Pioneer plaques, but rather to *Laverne and Shirley* and *Monday Night Football*.

The protests notwithstanding, most people seemed to feel that the Pioneer plaques had been a good idea. They excited interest in the Pioneer missions and presented a thoughtful rationale for the existence of extraterrestrial civilizations. They also, in a way, launched the career of Carl Sagan. As one of the creators of the plaques, he received numerous invitations to appear on television talk shows. Unlike many scientists, Sagan turned out to be an interesting TV personality. Articulate and enthusiastic, he used his television appearances to lobby for continued space exploration and thus became one of NASA's most valuable assets.

DISTANT ENCOUNTERS

In December, 1976, nine months before the first Voyager launch, Project Manager John Casani asked Sagan if he would like to coordinate an effort to include a Pioneer-style message on the Voyagers, since they, too, would ultimately leave the solar system. Since time was short and Sagan had been busy with his duties as a member of the Viking lander imaging team, his first thought was to send another plaque. But Frank Drake suggested that the message could be made much more durable and informative by sending it in the form of a phonograph record. This would enable them to include music, voices, and even pictures encoded as audio signals, much as is done on video discs.

Sagan was captivated by the idea and quickly set about organizing the creation of the record. He contacted scientists, philosophers, musicians, artists, and science-fiction writers, soliciting their suggestions about what the record should contain.

It was a tantalizing but very difficult challenge. After their experience with the Pioneer plaques, Sagan and Drake realized that the project had to proceed in relative quiet, lest they be inundated with proposals and demands by groups and individuals with particular axes to grind. At the same time, composing an interstellar message on behalf of all mankind could not be left to the whims and tastes of a small clique. They needed quick, quiet, broad-based input from experts in a wide variety of disciplines.

By May, Sagan had assembled a core group that would be responsible for composing and producing the record. They included Sagan, Drake, novelist Ann Druyan, *Rolling Stone* editor and journalist Timothy Ferris, space artist Jon Lomberg, and Linda Salzman Sagan.

The concept of the record was a lot like an intellectual party game: Which books would you take to a desert island, which musical recordings? Yet the record could not be just a collection of personal favorites and golden oldies. Ideally, it should contain a cross-section of the music of many cultures and times. The pictures ought to convey specific information about us and the planet we live on. And it all had to be comprehensible to a being who knew nothing at all about us.

The group had little more than six weeks to complete the project, an impossible deadline that they somehow met—in some cases, by mere minutes. Sagan coordinated the overall effort; Linda Sagan put together a collection of recorded greetings in as many languages as possible, Tim Ferris gathered the music and produced the recording "session," Ann Druyan assembled a "Sounds of Earth" essay, and Frank Drake and Jon Lomberg handled the pictures. In practice, there was considerable overlap of functions, and everyone ended up doing a bit of everything.

The Sagans, Ferris, and Druyan tended to see the record as a message *to* Earth, realistically recognizing that no one was ever likely to find it. Drake and Lomberg, however, were quite dedicated to creating a true interstellar message. "We were interested in making a message based on reaching extraterrestrials," says Lomberg, "a message *for* whoever found it." Lomberg's involvement in the project illustrates an important point about the Space Age: it's for everyone, not just highly trained scientists and engineers. Sagan has always been acutely aware of this and has spent years arguing, cajoling, and maneuvering NASA into expanding the base of its operations to tap the enthusiasm of the general public. Some NASA officials may have been a bit uneasy about their interstellar message being composed by artists and contributors to rock music magazines. But to Sagan, it was clear that the long-term success of the space program would depend on the support and participation of many kinds of people and not simply the technical wizardry of scientists and engineers. The proof was in the Voyager record itself, which was created by two scientists and four non-scientists.

The selection of music for the record stirred the most debate. Although Voyager and the record were the product of Western science and culture, the music had to reflect many other cultures. A number of people lobbied vigorously for their favorite composers, and there was a great deal of pro-Bach sentiment. Bach's highly structured, mathematically precise compositions would be ideal for an interstellar message because they could be enjoyed—or at least, understood—by someone who knew nothing about Earth. But with only ninety minutes of record time available for music, it would not have been practical to concentrate on the work of any one composer.

With the help of musicologists and record company executives, Ferris and the others put together a list of likely choices. Once the selections had been made, however, finding recordings of some of the songs proved to be a major problem. Bach and Beethoven were easy to come by, but recordings of Peruvian panpipes and pygmy initiation songs were not readily available. For a time, it looked as if some choices would have to be abandoned. Musicologist Robert Brown had strongly recommended an Indian raga, "Jaat Kahan Ho," but no one could find a copy of it. At the last moment, after some frantic detective work, Druyan managed to find a recording of the raga in an appliance store owned by an Indian family in New York City.

International politics posed other problems. Some NASA officials might have blanched at the thought, but the record team was determined to include at least one song from the Soviet Union. However, no one wanted to upset the Soviets by sending a counterrevolutionary song to the stars; whatever song was chosen, it had to be ideologically inoffensive. Everyone agreed on a Georgian chorus, "Tchakrulo," but the song was in Georgian and no one knew what the words meant. "For all we knew," wrote Ferris, "[it] could have celebrated bear-baiting." They managed to find a translator who assured them that the song was actually about peasants protesting against a landowner—a subject the Soviets could appreciate.

The worst problems were caused by copyright restrictions and record companies that wanted, somehow, to make money on the deal. There was talk that the Voyager record would ultimately be released as an album, which meant that legal technicalities had to be scrupulously observed. This resulted in the loss of the Beatles' "Here Comes the Sun." All four Beatles had given their permission for the use of the song, but the Beatles themselves did not control the copyright. The company that did control it refused permission, so the Beatles were not sent "across the universe." Rock 'n' roll was still well represented, however, by the inclusion of Chuck Berry's classic "Johnny B. Goode." Later, a *Saturday Night Live* sketch "reported" the aliens' response to the Voyager record: "Send more Chuck Berry."

In addition to the music, the record also contains spoken greetings in fifty-five languages that represent about eighty-seven percent of the population of Earth. The greetings were recorded at Cornell University, where speakers of many different languages were fortuitously available. In the end, only a few major languages were omitted, Swahili being the most prominent.

Some of the greetings were traditional, while a few were tailored for the presumed extraterrestrial audience. In Welsh, the message was universal: "Good health to you now and forever (Iechyd da i chwi yn awr ac yn oes oesoedd)." In Zulu, the greeting was humble: "We greet you, great ones. We wish you longevity. (Siya nibingelela maqhawe sinifisela inkonzo ende)." In Amoy, a language of Eastern China, the message was downright folksy: "Friends of space, how are you all? Have you eaten yet? Come visit us if you have the time." The Turkish greeting was wildly optimistic about what language would be spoken in the stars: "Dear Turkish-speaking friends, may the honors of the morning be upon your heads." The English greeting was recorded by young Nicholas Sagan: "Hello from the children of planet Earth."

When the news got out that an interstellar message was being prepared, it was inevitable that government officials would have their say. United Nations Secretary General Kurt Waldheim recorded a message ("We step out of our solar system into the universe seeking only peace and friendship, to teach if we are called upon, to be taught if we are fortunate."), although no one in the project had asked him to do so. Waldheim's unsolicited greeting had to be included, which meant that President Jimmy Carter would have to be asked to contribute something to the record. Carter elected not to record a message, but his typewritten greeting was included as one of the 118 photographs ("We are attempting to survive our time so we may live into yours. We hope someday, having solved the problems we face, to join a community of galactic civilizations."). And, of course, if the White House was represented on the record, Congress had to receive equal time. At the last minute, four additional pictures were included, these consisting of a list of Senators and Representatives who had approved funding for the Voyager mission. NASA knew who buttered its bread.

The recorded greetings reflect the "message *to* Earth" aspect of the record. For purposes of interstellar communication, it would have been preferable to send a message in just one language, along with some sort of Rosetta Stone that would enable the aliens to translate it. As it is, it seems unlikely that the extraterrestrials will be able to make much sense of our Earthly babel, no matter how clever they may be.

The aliens will not understand the words of our politicians (one trait, at least, that they will share with us), but they may be able to figure out the nonverbal "Sounds of Earth" essay on the record. The Earth noises are arranged chronologically and loosely depict our long evolutionary climb. Starting with the sounds of volcanoes, bubbling mud pots, and the crash of surf, the essay moves on to animal sounds, human speech, and the cacophony of modern technology, including a Saturn V lift-off.

Ann Druyan, who coordinated the sound essay, also sent the aliens a sample of her own thoughts. With the help of a doctor at New York University Medical Center, she hooked herself up to a medical data recorder which recorded her brain waves and other vital signs. Druyan spent an hour meditating, concentrating on ideas and images from history, hoping to give the extraterrestrials a glimpse of the best of humanity. The recording was compressed into a one-minute audio tape, which Druyan described as sounding "something like a string of exploding firecrackers." It is not inconceivable that technologically advanced aliens might be able to unscramble the EEG signal and actually "read" Druyan's mind. If they can perform such a feat, the one-minute tape will probably tell them more about us than all the rest of the record. Fortunately, Ann Druyan is an exceptionally intelligent and compassionate example of humanity; we are well represented.

The question of just how well we ought to present ourselves occasioned considerable debate. Should the record show us "warts and all," or should we send the extraterrestrials a kind of airbrushed graduation picture, all smiles? Was it possible to depict our technology without including military hardware? Should all the faces in the pictures be happy and well fed? Was it fair to send the recorded sounds of the humpbacked whale without somehow mentioning that mankind has driven that noble creature to near-extinction? In short, should we tell the truth about ourselves, or should we lie?

In the end, the record was, perhaps, a half-lie, or a half-truth. Weapons and explosions were vetoed on the very good grounds that the aliens might interpret such images as threats. The pictures chosen for the record do not reveal the seamier side of terrestrial existence, although a clever alien might be able to "read between the lines" and deduce some of the truth about us. The inclusion of so many different languages, for example, ought to be a clue that we are not a homogeneous, united planet.

Whatever the extraterrestrials may think of us, they are bound to be fascinated by the pictures on the record. Lomberg and a squad of picture hunters culled the 118 photographs from museums, libraries, private collections, and thousands of old magazines. Both the content and selection of the pictures reveal a great deal about us and the way we live our lives.

The pictures were chosen to depict specific information about us in the most unambiguous manner. Lomberg "played alien" and tried to look at each picture as if he had no prior knowledge of Earth. Things which were obvious to us might turn out to be complete mysteries to aliens. What would they make of a man riding a horse; would they think it was a centaur? Would they understand what clothing was? How would they interpret houses and airplanes and scuba divers?

Drake and Lomberg wanted the pictures to convey as much information about us as possible, individually and collectively. The picture sequence was designed to tell a story that would seem logical and comprehensible to beings with no knowledge of us or our planet. The sequence begins with some basic facts about Earth and the solar system, along with a rudimentary dictionary of signs and symbols that should enable extraterrestrials to understand our units of measurement and our counting system. To give the recipients of the message a qualitative understanding of us, it seemed essential that they should be given some quantitative yardsticks to work with.

Following some spacecraft pictures of the planets, the sequence moves on to diagrams of DNA and cellular structures. They should tell the recipients of the record that we are a carbon-based life form and that we reproduce by cellular division. Next, a set of eight pictures from *The World Book Encyclopedia* portray human anatomy. It will be an interesting exercise in comparative anatomy for the aliens to study these pictures, since the form of our organs doesn't necessarily reveal anything about their function.

The next sequence shows human reproduction, from microphotography of sperm and egg cells to the delivery of a baby. Fetuses and a pregnant woman are shown in silhouettes. There was to have been a photograph of a man and a pregnant woman standing nude, facing the camera, but after the controversy over the Pioneer 11 plaque, NASA vetoed the picture. We would be sending no more smut to the stars. No one was terribly surprised by the decision. "They did censor one picture," Lomberg says diffidently, "but they *didn't* censor 117 pictures."

After the reproduction sequence, the pictures become less scientific and more cultural in their orientation. By this point, the extraterrestrials ought to know *what* we are; the remaining pictures should tell them *who* we are. There are many pictures of children and families, which should tell the recipients something about the way we live and what we consider important. Other pictures show people at work and play, although it may be impossible for aliens to distinguish between the two.

The other residents of Earth are also well represented. The pictures show trees, flowers, insects, fish, toads, reptiles, birds, chimpanzees, and dolphins. Since the extraterrestrials might conceivably look like any of the above, or none of them, it seemed worthwhile to give them a look at the diversity of life on Earth.

After introducing ourselves and our planet, the pictures portray some of our accomplishments. There are several pictures showing different types of buildings, ranging from a thatched hut to the United Nations building and the Taj Mahal. Some of our machines are shown, with particular emphasis on transportation technology. But although the sequence shows carts and automobiles and trains and airplanes, the one vehicle which really ought to have been included is missing. There is no photograph of the Voyager spacecraft; despite many requests, NASA wasn't able to come up with a suitable picture in time to get it included on the record.

Near the end of the picture collection is a photo of a book by Sir Isaac Newton. Books and the written word have been essential to our civilization, so it seemed important to show the extraterrestrials how we communicate with ourselves. The photograph shows a thumb turning a page of the book; the thumb belongs to Lomberg. "A nice slice of immortality," he says.

The picture sequence concludes with a picture of a string quartet and a closeup of a musical score. The composition is the Cavatina from Beethoven's String Quartet Number 13, which is also the final

THE MESSAGE

musical selection on the record. Musical notation is quite logical, so it ought to be relatively easy for extraterrestrials to equate the symbols in the picture with the sounds in the recording. The picture also serves as a good introduction to the musical sequence which follows it.

The single greatest problem facing the record team was time. With less than six weeks to select, find, get permissions, and record the music and pictures, they inevitably had to settle for less than what they wanted. "We knew we could have made a much better choice in some areas than what we sent," said Lomberg. "We had to make some desperate choices."

The reaction of the people they contacted about the project was revealing. "Some people understood it before I stopped talking," says Lomberg. One man said it was a "screwy idea," and another hung up on him. "But nobody outright didn't believe me. To most people, it seemed reasonable. Many were thrilled and honored." Lomberg contacted one man about getting permission to use a picture of a baby being delivered. It turned out that the baby in the picture was, in fact, the man he was talking to, and permission was enthusiastically granted.

After many last minute crises, the completed recording in the form of bonded pairs of copper "mothers," one for each Voyager, was delivered to NASA and attached to the spacecraft at Cape Canaveral. The records were covered in an aluminum case, and a stylus and cartridge were attached nearby on the spacecraft. The diagrams on the record cover show how to use the equipment, and anyone clever enough to capture the Voyagers ought to be smart enough to figure out how to play the record.

The participants in the record project gathered at Cape Canaveral on August 20, 1977, for the first Voyager launch. It was an emotional experience for all of them. "I don't really believe in astral projection and that stuff," said Lomberg, "but if it's really possible to focus all of one's attention on a single spot, I felt like I was part of the launch, rather than just observing the launch." For Lomberg, a science-fiction devotee, the experience seemed to intermingle the best of reality and fantasy. "There was a sense that it would never be coming back . . . it would be crossing the galaxy. This was real. This wasn't a story."

The ultimate fate of the Voyager interstellar record will never be known. The flight paths of the two spacecraft will probably never take them within billions of miles of another solar system, although Sagan points out that it might just be possible to give Voyager 2 one final course correction that would send it very close to the star AC+79 3888 some 60,000 years from now.

There are differing schools of thought about what would be the best fate for the Voyagers. Science-fiction author Robert A. Heinlein suggested that each spacecraft be fitted with a small radar reflector, which will make it easier for our descendants to track down the spacecraft, retrieve them, and put them in a museum. Arthur C. Clarke, on the other hand, urged that the record contain a message to those space-faring descendants: "Please leave me alone; let me go to the stars."

Lomberg agrees with Clarke. "I think about the range of alternatives a lot," he says. His least favorite alternative is that the Voyagers should wind up in the Smithsonian 500 years from now. He would also be disappointed if the spacecraft eventually stumbled into a permanent orbit around an uninhabited star system.

But Lomberg, like the rest of us, will never know. That is inherent in the business of putting notes in bottles, but still we do it. If an extraterrestrial ever does find Voyager and the record, he (or it) will realize that the beings who launched it had no way of knowing if it would ever be found. Yet it will be clear that those odd-looking creatures on the third planet of a nondescript star devoted a great deal of thought and effort to creating this message. Whoever they were, they cared very much about the future.

5 — FIRST ENCOUNTER

There had never been anything like it. For two weeks in late February and early March, 1979, Voyager 1 plunged through the Jovian system, shattering theories and changing forever the way in which earthlings look at the universe. The high-tech, soberly scientific Voyager mission turned into something different, something more—it was an interplanetary freak show, an expedition to the other side of the looking glass, where the Merry Prankster Imaging Team provided the pictures and Lewis Carroll explained the science. Blindfolded scientists whacked the cosmic piñata and were astonished by the goodies that tumbled out of it. Magnificent, majestic Jupiter, king of Olympus, sultan of the solar system, grand Poo-bah of the planets, at last revealed its true DayGlo colors; it was like going to the Milan Opera House and finding the Grateful Dead on stage, or to the Louvre and seeing the cover of Zap Comix in place of the Mona Lisa. Jupiter—the psychedelic planet.

It began with a routine roar of rocketry, as Voyager 2 lifted off from Cape Canaveral on August 20, 1977. Voyager 1 was scheduled for departure on September 1. The second Voyager went first because of the carefully planned trajectory for Jupiter and Saturn encounters. In effect, Voyager 2 was a big, roundhouse curve ball, while Voyager 1 was a fast ball. By December, Voyager 1 caught up with and passed Voyager 2 and beat it to Jupiter by about five months.

Shortly after launch, Voyager 2 began displaying the temperamental behavior that was to typify its entire flight. Signals from the spacecraft indicated that the science scan platform boom had not deployed properly from its stowed launch position. Engineers toyed with the problem for several days and finally concluded that the boom was in correct position after all, and that the fault was in the electronic sensor monitoring the boom's position.

Ironically, Voyager 2's problems indirectly benefited Voyager 1. The Voyager 1 launch was scheduled for September 1, but with everyone concentrating on the Voyager 2 science boom, it was decided to delay the second launch until September 5. The delay meant a slight change in the planned trajectory and the timing of a midcourse burn for the Titan encounter at Saturn; happily, the altered trajectory would require a smaller change in velocity than the original. By launching four days late, Voyager 1 "saved" about 40 pounds of hydrazine fuel, which could come in handy if unexpected problems arose later in the mission.

For Voyager 2, however, the amount of hydrazine on board was critical if there was to be a Uranus encounter. At launch, Voyager 2 had a hydrazine margin for Uranus of about 15 pounds. But it was soon discovered that the thrusters were delivering about 20 percent less power than they were supposed to. Since more fuel would be needed to compensate for this use, the extra 15 pounds of fuel effectively vanished. Engineers eventually determined that the plume from one of the thrusters was hitting part of the spacecraft, reducing its efficiency. For Voyager 1, the problem was not terribly important; for Voyager 2, it was a potentially lethal threat to the Uranus option.

Charlie Kohlhase described the Voyager 2 hydrazine story as "a soap opera." From a prelaunch margin of +15 pounds, in-flight events and glitches lowered the hydrazine level to −30 pounds at one point. But as the mission controllers learned more about their spacecraft ("It took us over a year to learn how to fly these things," said

Deputy Project Manager Esker "Ek" Davis) the effective hydrazine margin was increased. By "fine-tuning" the trajectory and finding the most fuel-efficient in-flight attitude, the hydrazine margin was returned to positive numbers. The Uranus option was saved.

There remained a question as to whether the controllers would be able to communicate with the spacecraft if it did go on to Uranus. On April 5, 1978, Voyager 2's primary radio receiver failed, as did the secondary receiver. The failure was due to the most common of human errors—forgetfulness. The computer on Voyager 2 was programmed to switch to the backup receiver if more than a week went by without instructions from Earth. Incredibly, the controllers simply forgot to keep in touch with their spacecraft, so the computer executed the failure routine.

Unfortunately, the secondary receiver had a glitch in its tracking-loop capacitor. The frequency of signals from Earth to a spacecraft changes due to the Doppler effect—both the spacecraft and the Earth are moving rapidly, shifting the frequency of the radio signals between them. Without the capacitor, Voyager 2 could not lock onto the signals from Earth. After twelve hours passed with no new instructions, the failure routine required that the spacecraft computer switch back to the primary receiver—which promptly blew a fuse. With the primary receiver gone for good, the engineers had to find a way to get the secondary to lock onto the Doppler-shifted radio signals; otherwise, the entire Voyager 2 mission would be a failure.

The solution was to calculate the frequency at which the radio signals would be arriving at Voyager 2, and then change the broadcast frequency to compensate for the Doppler effect. That seems simple enough, but it meant completely revising the uplink procedures. Fortunately, the three main antennas of the Deep Space Network could be programmed to vary the uplink frequency, although this was a contingency no one had planned for. The procedure worked, and as long as the secondary receiver remained functional, Voyager 2 would stay alive.

Other problems arose during the cruise phase. It was during this phase that the photopolarimeters malfunctioned and were effectively lost for the Jupiter encounters. At one point, the Voyager 1 scan platform stuck, and the engineers had another exercise in long-range diagnosis and repair. This glitch eventually solved itself, and the

Voyager 1 Flyby of Jupiter March 3–6, 1979

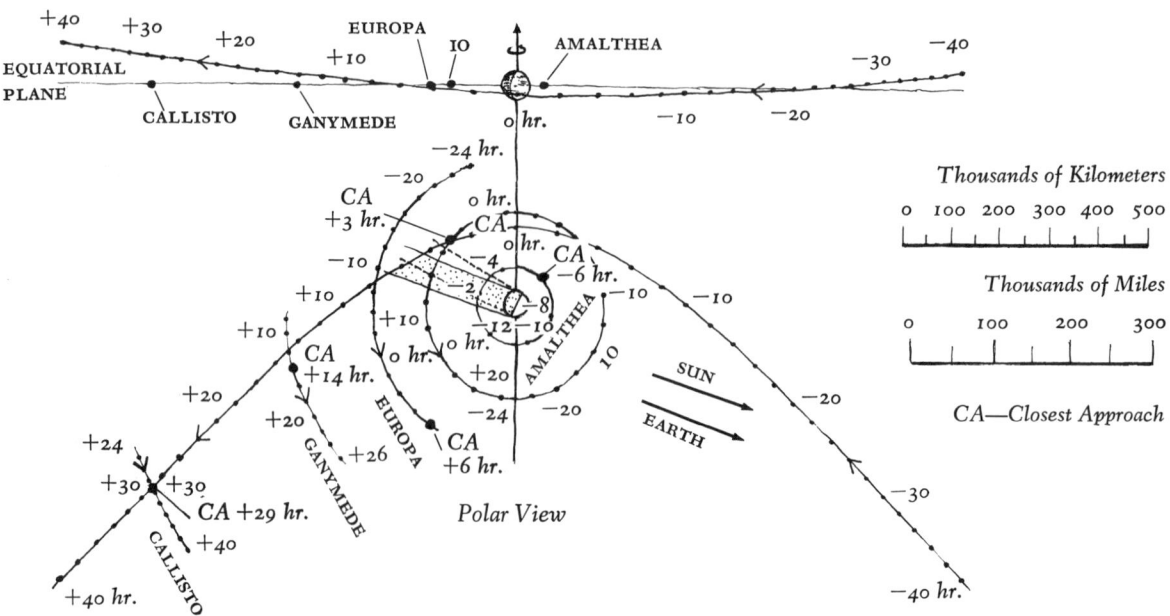

By following the arrows, you can see at what points in its trajectory Voyager 1 made its closest approach to Jupiter and its satellites. The top portion of the diagram looks at the flyby edge on; the bottom portion is looking down on the Jovian system.

theory was that a bit of foreign matter had lodged in the gears of the scan platform; the engineers presumed that as the scan platform maneuvered, it crushed or knocked aside the extraneous material.

One by one, the glitches came and went, were solved or not, and the Voyagers sailed onward. By the end of 1978, Voyager 1 approached Jupiter in good health. The show was about to begin.

Pioneers 10 and 11 had already given us a preview of Jupiter with their flybys in 1973 and 1974. Smaller and less sophisticated than the Voyagers, the Pioneers showed that it was possible for a spacecraft to cross the asteroid belt, pass through the high-radiation environment of Jupiter, and survive. They also returned valuable scientific data on the gas giant.

Unlike the Voyagers, the Pioneers were not spin-stabilized. They rotated rapidly around a central axis that was always pointed at the Earth. This permitted them to employ a simpler guidance system, but it also limited the effectiveness of the Pioneer imaging system. Each picture had to be put together line by line, one line per spacecraft rotation. The resulting images were three times better than the best ground-based photographs of Jupiter, but were far less detailed than the images from Voyager would be.

Pioneer 10 passed within 3 Jupiter radii (R_J) of the center of

FIRST ENCOUNTER

the planet, or 2 R_J (140,000 km) above the visible cloud tops.*
Pioneer 11 came even closer, just 34,000 km from the topmost
Jovian clouds. Both spacecraft spent several days inside Jupiter's intense radiation field, and both were affected by it. The radiation caused the computers to issue spurious commands to the imaging systems and the Pioneer 11 infrared radiometer. Some science data were lost due to the malfunctions, but the important point was that both spacecraft were still functional after emerging from the Jovian radiation.

The Pioneers gave us our first *in situ* measurements of the Jovian environment. They found that the effective global temperature at the cloud tops was $-148°C$, and that the temperature did not vary significantly between the equator and the poles. The presence of an internal heat source was verified, and its total energy output was calculated at 10^{17} watts. Helium in the Jovian atmosphere was measured for the first time, and a magnetic field 2,000 times stronger than that of the Earth was detected.

The Pioneer results were important, but they didn't exactly startle anyone. In the best Pioneer images, with a resolution of about 500 km, Jupiter does not look dramatically different than it does through a large telescope. The Pioneer imaging system was simply unable to show the fine structures of the Jovian clouds, so the details of Jovian atmospheric circulation remained a mystery. As for the Galilean satellites, the Pioneer images were fuzzy and showed about as much detail as the Earth's moon when seen through a cloudy sky. The Pioneers had done their jobs remarkably well, but they didn't prepare anyone for the wonders Voyager was about to unveil.

Voyager 1 entered its observatory phase on January 4, 1979, at a distance of more than 50 million km from Jupiter. Even at that range, the scientists gathered at the Jet Propulsion Laboratory in Pasadena, California, could tell that the Jupiter they were about to encounter was strangely different from the Jupiter encountered five

---

*For encounters with the gas giants, planetary radii are the most useful yardstick for measuring the spacecraft's distance to the planet. Although the radius of Jupiter is some 10,000 km greater than that of Saturn, a spacecraft 4 R_J from Jupiter would be at the same relative distance to the planet as a spacecraft 4 R_S away from Saturn. The use of radii makes it easier to compare data from the Jupiter and Saturn encounters.

years earlier by Pioneers 10 and 11. Ground-based astronomers had reported seeing increased turbulence and activity in the Jovian atmosphere, and now Voyager confirmed the view from Earth. The Pioneers had seen a bland, calm, muted Jupiter; Voyager was seeing a swirling, tempestuous, riotous Jupiter. The improved imaging system of Voyager would permit the scientists to see more, but there was also more to be seen.

Throughout January, Voyager took photographs of Jupiter every two hours. Although the smallest discrete object visible in these images was about the size of Asia, each passing day revealed new details about the Jovian atmosphere. Attention naturally focused on the Great Red Spot. Since the 1960's it had been known that the spot was rotating in an anticyclonic (counterclockwise) direction once every six days. On Earth, anticyclonic rotation is characteristic of high-pressure systems, rather than of low-pressure hurricanes; thus, there was considerable uncertainty about the validity of the hurricane

An early mosaic of Jupiter, assembled from nine individual images taken from a range of 4.7 million miles. Scientists were astonished by their first views of Jupiter's riotous atmosphere. The Great Red Spot is at lower right. The regular black dots are calibration marks on the camera itself; they appear frequently in Voyager images.

analogy. Intensive observation by Voyager, it was hoped, might lead to a better understanding of the spot's nature.

By the beginning of February, Voyager 1 entered the far-encounter phase. Pictures were taken every 96 seconds for a period of 100 hours. The resulting images produced the equivalent of a movie, showing the planet's rapid rotation and swirling belts and zones. The activity around the Great Red Spot was almost beyond belief. Material in the horizontal belts was sucked into the vortex of the spot at speeds greater than 100 meters per second. Some of it stayed in the spot's rotation system for hours or days; some material was quickly ejected to the east or west of the spot. Small bright and dark oval-shaped features that encountered the spot would break up, and then somehow reform as they spun out of the spot, back into the zonal flow.

The movie also revealed that the horizontal flows were not limited to the equatorial regions. Earth-based observers could not see the polar regions, but following the Pioneer encounters, scientists theorized that the zonal flow would break down at high latitudes and be replaced by polar vortices. Instead, Voyager showed horizontal flows extending far to the north and south.

By the end of February, Imaging Team Leader Brad Smith described himself and his colleagues as "happily bewildered." The Jovian atmosphere, "where our greatest state of confusion seems to exist," was not behaving the way everyone expected. The neat and orderly drawings of belts and zones suddenly seemed irrelevant. "Pioneer gave us snapshots of Jupiter," Smith pointed out, "but told us nothing about the motions of the atmosphere. I think, for the most part, we have to say that the existing atmospheric circulation models have all been shot to hell by Voyager. Although these models can still explain some of the coarse zonal flow patterns, they fail entirely in explaining the detailed behavior that Voyager is now revealing."

Smith compared the Pioneer images to a still picture of a football game. "It can see a lot of action out there, but it can't tell us who's winning. Voyager will tell us who's winning."

Clearly, Jupiter was winning, knocking the theoretical stuffings out of the scientists. The Jovian atmosphere was running a razzle-dazzle offense that left the scientists gawking in amazement. Smith described the behavior of some small (for Jupiter), swirling filamentary cloud structures that collided with the rotation system of the Great Red Spot. At first, they "looked like a couple of twined-up pieces of spaghetti," but as they spun around the spot they diffused into amorphous blobs. Coming out of the spot system, they somehow maintained their integrity and reformed into filaments. "We have no explanation for that type of behavior," Smith said.

The rest of the world was beginning to share in the excitement. By the beginning of March, the press corps had descended upon the Jet Propulsion Laboratory like a swarm of ravenous locusts. Daily press briefings were held in Von Karman Auditorium, and scientists were called upon to provide "instant analysis" of the Voyager data. Besides the formal news conferences, there were many casual briefing sessions, plus an ongoing television "play by play" as events unfolded. The television operation, known as the "Blue Room circuit," was hosted by Dr. Al Hibbs, who had performed the same duty during the Viking mission. Hibbs provided informative interviews with the scientists, but the press corps' thirst for news was insatiable. Any scientist who ventured into Von Karman was likely to find himself waylaid by the hungry mob.

The scientists, too, felt the excitement. After dreaming of this encounter for years—in some cases, for their entire professional lives—they found the reality almost more than they could handle. "It may sound unprofessional," Brad Smith reported, "but a lot of people up in the imaging team area are just standing around with their mouths hanging open, watching the pictures come in, and you don't like to tear yourself away to go and start looking at numbers on a printout. We'll do that, but in the meantime we're just caught up in the excitement of what's going on."

A lot of mouths were hanging open around JPL that week. Television monitors all over the Lab displayed the latest Voyager pictures as they were received. The raw images were sometimes difficult to appreciate or interpret, but no one could resist staring at the monitors and trying to figure out the many mysteries of Jupiter. In the press room, some monitors were hooked into the interactive loop, which displayed computer-processed images from the Image Processing Laboratory (IPL). The processed pictures included color images that were nothing short of stunning.

FIRST ENCOUNTER

No one had ever seen pictures like these before. Whatever the precise scientific explanations might turn out to be, the mere fact that such scenes existed was mind-boggling. The interplay of the Jovian cloud formations, with their delicate filaments, roiling whirlpools, and subtle colors, created a kinetic, ever-changing, utterly unique art form: "a new kind of design concept," said Jon Lomberg. As an astronomical artist, Lomberg noted that "none of the great space artists got it right. . . . Jupiter is the most alien place we've seen." Lomberg's personal reaction summed up the scene about as well as anyone's. On being shown one of the first of the full-globe cloud pattern pictures, he jabbed a finger at the photograph and declared, "That planet's on *acid!*"

While the imaging team stared at the approaching clouds, the particles and fields experimenters waited for their instruments to signal news of Voyager's arrival at the Jovian magnetosphere. At the outer boundary of the magnetosphere, known as the bowshock, the solar wind breaks like a wave hitting a reef. The speed of the solar particles suddenly drops from nearly a million miles per hour to about 250,000 mph. Between the bowshock and the magnetosphere proper lies a region known as the magnetopause. Pioneer 10 first encountered the magnetopause at 109 Jupiter radii from the planet; Pioneer 11 found it at 110 R_J. But by the end of February, Voyager 1 was less than 90 R_J from the planet, and there was still no sign of the bowshock.

Millions of miles closer to the sun ("upstream"), Voyager 2 was sending back data on the solar wind pressure that indicated a stronger than normal flow of particles from the sun. The Jovian magnetosphere was being compressed by the solar wind. This phenomenon had been observed during the Pioneer missions, as the fluctuating strength of the solar wind caused the magnetosphere to expand, contract, and flap around like a windsock in a hurricane. The Pioneers crossed the bowshock several times as the magnetosphere pulsated back and forth, in and out.

Voyager 1 finally hit the bowshock on February 28 at a distance of 86 R_J. The solar wind strength then increased, the magnetosphere was compressed, and at a distance of 82 R_J, the spacecraft again crossed (or was crossed by) the bowshock and found itself back in

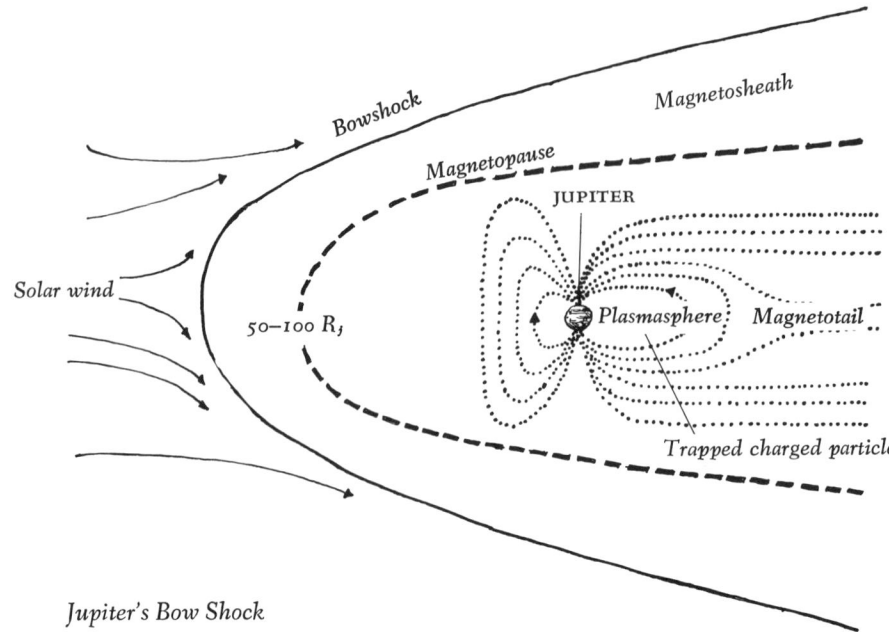

Where the bowshock is located at any given time depends on the strength of the solar wind. Jupiter's magnetic field is compressed on the side toward the sun, concentrating the radiation through which Voyager was to pass.

Jupiter's Bow Shock

the interplanetary medium. The next day, Voyager 1 caught up with the bowshock and recrossed it at 72 R_J. A few hours later, the spacecraft crossed the magnetopause and entered the magnetosphere proper at 67 R_J. But the field contracted again, and on March 2 the magnetopause swept past Voyager at 59 R_J, followed by the bowshock at 58 R_J—Voyager was once again back in the solar wind. The spacecraft crossed the bowshock for the fifth and final time on the inbound leg of the encounter at a distance of 56 R_J. On March 3, at 47 R_J, the magnetopause was crossed again, and Voyager was at last within the magnetosphere to stay.

The compression of the magnetosphere was a cause for concern. Would the radiation levels inside the magnetosphere be concentrated beyond the levels observed by the Pioneers? If so, the spacecraft might sustain more radiation damage than had been anticipated.

The environment inside the magnetosphere was turning out to be wildly different from the expectations of the physicists. Donald Shemansky reported "spectacular results" for the ultraviolet spectroscopy team on March 2. Previously, the Io torus—the doughnut-shaped cloud of ionized gases surrounding the orbit of Io—was thought to consist mainly of sodium, with some potassium and possibly some hydrogen. But, Shemansky said, "we were surprised out of our chairs to see a spectacularly bright emission in the 650 to 1,100 Ångstrom region, immediately implying that we were looking at a plasma that had to be at a temperature of 100,000 degrees." The emission meant that some of the torus was composed of doubly ionized sulfur—sulfur atoms in which two electrons had been stripped away by high-energy interactions within the magnetosphere. Since sulfur in that form doesn't survive for long before recombining with other electrons, and since some sulfur is lost from the Jovian system, the torus had to be constantly replenished with more sulfur, presumably coming directly from the surface of Io. The energy required to maintain the torus was enormous—on the order of 500 billion watts, or about a thousand times greater than anyone expected.

The torus was found to be rotating in synchronization with Jupiter's magnetic field, rather than with Io. The moon moved in and out of the torus as it traveled its orbital path. The surface of the moon, it seemed, was being bombarded by high-speed subatomic particles from the magnetosphere, kicking up sulfur and oxygen, in the form of sulfur dioxide. Some of the sulfur remained in a cloud moving with Io, but most of it was quickly grabbed by the magnetic field and locked into the torus. In the torus, the sulfur was somehow energized, but no one was sure precisely how this was accomplished. Ultraviolet Spectroscopy Team Leader Lyle Broadfoot commented, "The physics of the planet is obviously far different from what we had anticipated." For the particles and fields men, all of this was as exciting and bewildering as the Jovian clouds were for the imaging team.

Jupiter and the magnetosphere were dominating the stage, but waiting in the wings were the Galilean satellites. Voyager 1 would make its closest pass to Io, Ganymede, and Callisto on the outbound leg of the encounter, following the nearest approach to Jupiter. Europa would be seen only from a great distance, and closeups of that moon would have to wait for Voyager 2. There was also a chance that images would be obtained of the innermost moon, Amalthea. No one could guarantee images of Amalthea, however, because no one was precisely sure where it was. University of Hawaii astronomer David Morrison confessed, "I have never even succeeded in *seeing* Amalthea," even though he was an experienced observer using some of the world's best telescopes. Voyager was programmed to aim its cameras toward the spot where everyone hoped Amalthea would be; but if the moon was only slightly out of position, the picture would be missed.

A few spectacular long-range pictures of the Galilean satellites had already been obtained. A color picture of Jupiter taken on February 13 at a distance of 20 million km showed a vivid orange Io floating above the Great Red Spot, with the frosty, white sphere of Europa nearby. The combination of color and composition made this one of the most beautiful and artistic images ever received from a spacecraft. Other images of the Jovian clouds happened to catch Io in the frame, looking like a small, pitted marble rolling around on a sheet of rococo linoleum. Voyager was still too far away for the pictures to show much detail, but it was becoming clear that the surfaces of the Galilean satellites were unlike anything ever seen in the solar system.

At the start of the encounter, the geologists on the imaging team found a certain amount of amusement in the bewilderment of the Jupiter experts. On February 28, after admitting his confusion over the behavior of the Jovian atmosphere, Brad Smith noted that "Over the next several days we may find that some of our smirking geology friends will find themselves in a similar state." One of the smirking

Io intruded on this segment of a Jupiter mosaic. Even from 5 million miles away, the tiny moon looked strange.

geologists was Deputy Team Leader Laurence Soderblom from the U.S. Geological Survey at Flagstaff, Arizona. Soderblom, a veteran of the Mariner and Viking missions to Mars, compared the coming satellite encounters to the historical growth of our knowledge about Mars. "It is about 1700 A.D. this morning," he said on Thursday, March 1; "tomorrow it will be about 1800, and it will be about 1975 by Tuesday evening."

"We're beginning," said Soderblom, "a stage in this mission which represents, I think, one of the most exciting points in man's scientific exploration of the solar system. In the next few days, we'll explore four new worlds." Soderblom's use of the word "worlds" to describe the Galilean satellites was revealing. Calling these bodies "moons" was somehow no longer appropriate. Our experience with moons had been limited to our own battered satellite and the two tiny, misshapen lumps of rock orbiting Mars. These bodies had their intrinsic interest, but they did not have the diversity and complexity that was becoming apparent on the Galilean satellites—the Galilean *worlds*. More than one observer would note during the encounter week that seeing Io, Europa, Ganymede, and Callisto for the first time was like exploring a brand-new solar system—something none expected to happen in their lifetimes.

The four new worlds (or planets, as everyone now thought of them) had never been seen as anything but blurred points of light, even in the largest ground-based telescopes. Many people expected that they would look a lot like the Earth's moon or Mercury, with surfaces pockmarked by thousands of impact craters. During the last stages of planetary formation, four and a half billion years ago, unconsolidated debris crashed down onto the surfaces of the young planets. Earth, with a dense atmosphere and an active geology, covered up or eroded most of those craters. Mars, with a thinner atmosphere and a less active interior, only partially erased the scars of the ancient bombardment. Mercury and the Earth's moon, with no atmospheres, preserved the cratering record. Since none of the Galilean satellites have atmospheres, and since they were thought to be too small for radioactive heating to be able to drive an active surface geology, it was reasonable to suppose that their surfaces would resemble those of the Earth's moon and Mercury. On the other hand, the outer solar system was a completely new environment to us, and the history of the Jovian system may have been quite different than that of the inner solar system. Everyone expected ancient surfaces dominated by craters, but there was considerable disagreement about what craters would look like on icy bodies such as Ganymede and Callisto.

The first long-range images of Callisto seemed to confirm the Mercury analogy. It looked as if there were a great many impact craters on the surface of Callisto. Yet Callisto is a low-density, icy world, unlike the dense, rocky Mercury. Geologist Hal Masursky thought it "extraordinary" that Callisto should look like Mercury, since the two worlds "are the most unlike things in the solar system." A few days before closest encounter, Masursky suggested that it was possible that "we're being fooled."

Moon analogies were also suspect. The Earth's moon is dark and dense, with an albedo of 12 percent (that is, the surface of the moon reflects 12 percent of the light that hits it) and a density of 3.5 grams per cubic centimeter. Io, innermost of the Galilean worlds, has the same density as our moon, but an albedo of 60 percent. Next out, Europa has an albedo of 65 percent but a slightly lower density. Ganymede's albedo is 45 percent, while Callisto's is only 18; both of the outer moons have densities of about 2.0, twice as dense as ice but much lower than ordinary rocks. Like the Earth's moon, all four Jovian bodies are in tidal lock with the planet, presenting the same face to Jupiter at all times. But while Io's leading hemisphere is brighter than its trailing hemisphere, on Callisto just the reverse is true. Europa and Ganymede have the same sort of asymmetry, but rotated by 60 and 120 degrees, respectively. This systematic variation was another source of puzzlement for the astronomers.

All four bodies have a rocky mass that is about equal to that of the Earth's moon. Io and Europa are about the same size and density as it is; Ganymede and Callisto are much larger, implying that the excess mass consists of ice. If the young Jupiter radiated heat, the water would have been "cooked" out of the two inner moons. However, the extremely high albedo of Europa implied a brightly reflective icy surface. Callisto, on the other hand, should have consisted of a large amount of ice; but its surface was the darkest of the four. Somehow, dark material had covered the ice. On Ganymede, Voyager was beginning to reveal long bright streaks that may have been subsurface ice thrown up by impacts; but why weren't there similar bright streaks on Callisto?

Each new scrap of information raised new questions. The geologists were pressed for answers, but no one was brave enough or foolish enough to attempt detailed explanations only a few days before closest approach. For the most part, they were happy just to sit back and enjoy the show. Dave Morrison of the imaging team summed up the prevailing attitude toward the Galilean satellites: "You can't imagine how exciting it is to be seeing them as worlds for the first time."

The Jupiter specialists shared Morrison's emotions. The images of the Jovian atmosphere were revealing more detail each hour. "The excitement to me," said Andy Ingersoll, "is that we see so much turbulence, so much transience; yet we know that many of these flow patterns last for hundreds of years." Said Garry Hunt, an imaging team member from University College in London, "We didn't expect big changes four rotations apart—and we find terrific changes in even one rotation!" The fine details of atmospheric circulation were baffling everyone. "When Brad says everything is shot to hell," said Ingersoll, "he may just be seeing Jovian *weather* for the first time, instead of Jovian *climate*."

As they watched the Voyager images arrive, the atmospheric scientists knew that they would have a formidable task ahead of them. To explain the Jovian atmosphere, they would have to come up with a theory which encompassed the east–west flows in the belts and zones, the small white and dark blobs, the sawtoothed patterns that separated some of the belts and zones, the "hot spots" revealed by holes in the upper cloud layer, *and* the Great Red Spot. Without the Spot, things might have been easier, but it was impossible to ignore. Was it a hurricane, or a terrestrial "blocking high," or was it something uniquely Jovian, without Earthly parallel? Hurricanes last for weeks, blocking high-pressure systems endure for months; but the Spot had been around for hundreds, possibly millions, of years. Why did it exist at all? And why was it red?

What the particles and fields teams found was not so obvious or dramatic as a color picture of the Great Red Spot, nor was it as interesting to the press corps and the general public. Their only "media event" was plasma-wave experimenter Fred Scarf's tape of

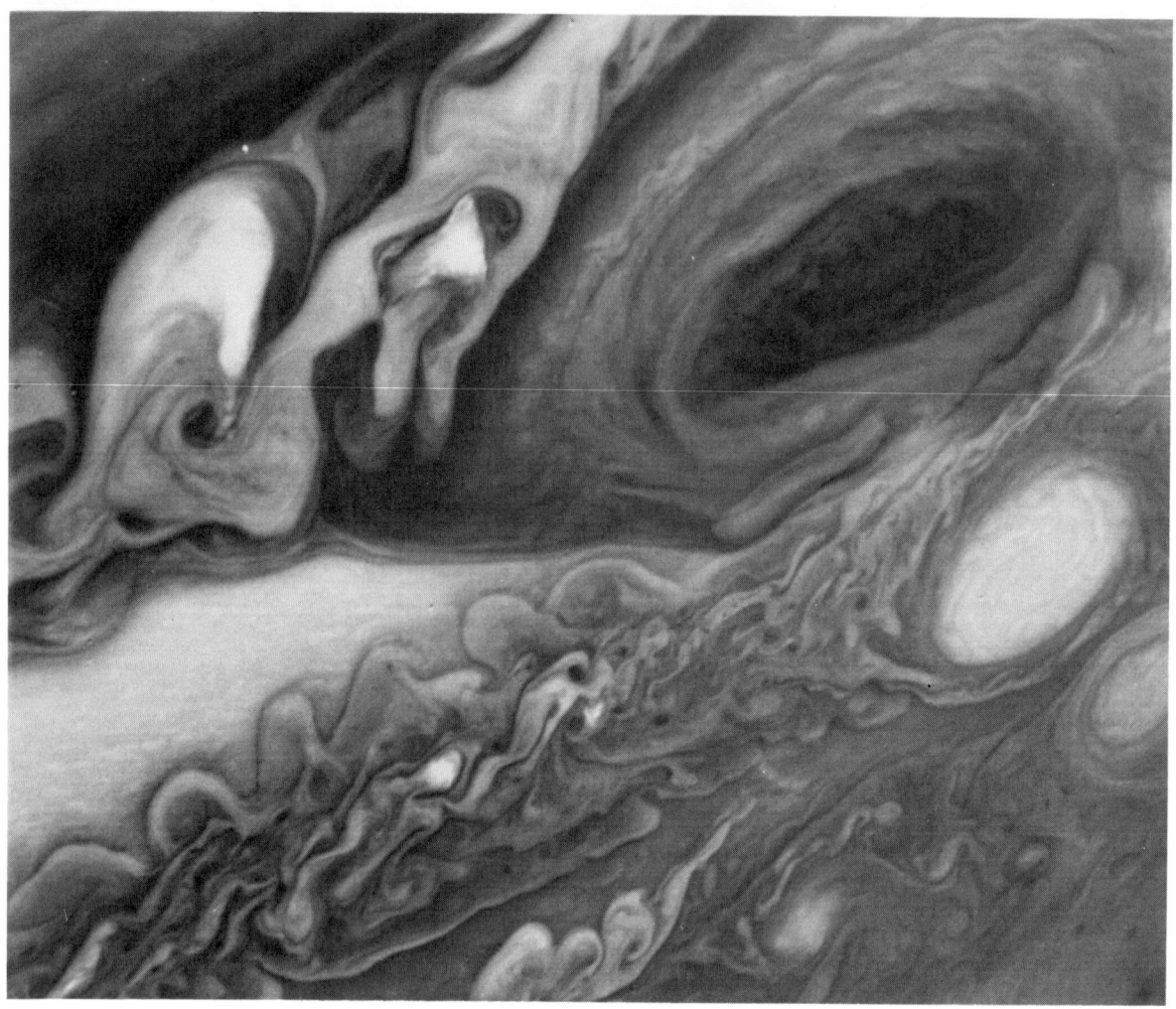

the sounds of Jupiter. The plasma-wave antennas picked up radio noise from protons as they zipped "upstream" through the bowshock. The antennas also picked up random noises from the spacecraft itself. The resulting tape, which Scarf played at a press conference, sounded like a catfight in a dentist's office.

Voyager 1 would make its closest pass to Jupiter and Io during Sunday, March 4, and the early morning hours of Monday, March 5. The spacecraft was still performing well, despite its exposure to the Jovian radiation. What mattered was the total radiation dose the spacecraft received during the encounter, a product of the intensity of the radiation and the length of the exposure. Negative effects would probably not become apparent until after the closest encounter with Jupiter.

Back on Earth, another sort of problem was making itself felt. Bad weather over the Deep Space Network stations in Madrid and

The atmospheric turbulence around the Great Red Spot, seen in detail for the first time. The smaller of the two ovals to the right (partially cropped) of the Spot is about the size of the Earth.

Australia had interfered with the incoming signals from Voyager, and several hours' worth of data had been lost. On March 2, an intense thunderstorm in Australia seemed to sit right over the station for three and a half hours. Some important Io observations were lost, just as imaging of the Great Red Spot had been ruined by an earlier storm. Terrestrial weather was something no one had control over, and the Voyager engineers could only keep their fingers crossed and hope for fair skies during closest approach.

The weather in Pasadena was excellent as scientists and reporters assembled in Von Karman Auditorium for the daily press briefing at 11 AM, Sunday, March 4. Everyone seemed a little giddy that morning, like football players on Superbowl morning. This was the beginning of the Big Day, the closest approach to both Io and Jupiter, and most people would be at JPL continuously for the next thirty-six hours. As Larry Soderblom put it, "Today is probably going to turn out to be one of the most memorable days in the history of planetary exploration. It's truly Christmas Eve for the planetary geologists. We see tonight the beginning of the exploration of four new worlds. We're racing through time and space at an incredible rate—the rate at which we're learning things is awe-inspiring in itself."

Soderblom showed the press the latest images of the Galilean satellites. Callisto was still too far away for detailed pictures, but many small bright spots indicative of cratering were becoming apparent. The geologists were eager to see closeups because cratering on icy Callisto might look very different from cratering on the Earth's moon or Mercury. No one was sure how the ice would react to large impacts. As Soderblom had put it two days earlier, "When you take a sledge hammer to somebody's car windshield, it doesn't melt."

Next, Soderblom showed the latest Ganymede pictures. "What can I say?" he asked. There were bright streaks and "odd-shaped features . . . loops and swirls and incredible patches that are difficult to hazard a guess about." If Callisto resembled Mercury, Ganymede was beginning to look like a slapdash planet hastily assembled from spare parts.

Europa, seen from two million kilometers away, was still a bland and craterless white orb; closer inspection would have to wait for Voyager 2. But the Voyager 1 images did reveal some very fas-

cinating dark streaks. If the streaks turned out to be valleys, they would be comparable to the huge Valles Marineris on Mars.

Soderblom saved the best for last. He stood on the Von Karman stage next to the large movie screen and grinned as the projector flashed the latest images of Io. "Holy Jesus!" cried out someone in the audience. The picture revealed a vivid orange, red, and yellow sphere with an immense heart-shaped (or hoofprint-shaped) feature domi-

From 2.6 million miles, Ganymede (left) looked deceptively similar to Earth's moon. Europa (right), from 1.75 million miles, first revealed its intriguing dark streaks.

The "Eye of Io" was a baffling, unexpected sight. This hoofprint-shaped feature, 600 miles across, intensified speculation about Io's odd surface. Where were all the impact craters?

nating the southern hemisphere. "This one," said Soderblom, "we've got all figured out."

This bizarre "eye" of Io* was about 800 by 1,000 km in size. A ridge about 100 km wide seemed to surround the feature, with some kind of large central peak in the middle of the hoofprint. If this was an impact crater, it looked like no other ever seen in the solar system. However, the total absence of other obvious impact craters made any interpretation of the hoofprint very speculative. Not much could be said about it yet. "Whatever that structure is," said Dave Morrison, "it's made of the same surface material that's on the poles . . . what that means, I have no idea."

While Soderblom and company were wowing the press corps, the spacecraft was crossing Jupiter's equatorial plane. The trajectory carried the spacecraft on a long, curving parabola from north to south across the Jovian equator. Years earlier, Voyager had been programmed to take a single, long-exposure image as it passed the equatorial plane; since there was some uncertainty about the exact position of the spacecraft (and of Jupiter), the exposure was set for eleven minutes. The narrow-angle camera was aimed for a point in space about midway between the orbit of Amalthea and the Jovian cloud tops. The picture was literally a shot in the dark. Its purpose was to find a hypothetical ring around Jupiter.

The ring wasn't hypothetical. It was there, and the image caught it dead center, as if the scientists had known all along that a Jovian ring existed. The time exposure didn't look like much, and when it appeared on the TV monitors, most people simply ignored it. Even the scientists were not convinced that they really had the ring. News of the discovery was circulating in the press room within hours, but no formal announcement was made until three days later, on Wednesday, the 7th.

Sunday afternoon, following the press briefing, another long shot paid off. The preprogrammed picture sequence captured an image of tiny Amalthea, despite the uncertainty about its location. The image was only about 35 pixels across, and no details would be

---

* The correct pronunciation of Io is "Ee-oh," according to some experts. Most people at JPL, including some scientists, called it "Eye-oh." A few seemed to alternate between the two. The name was troublesome for everyone. One newspaper account referred to it as "the moon, Ten."

visible until the picture was computer-processed. But it was clear that Amalthea is more like the miniscule moons of Mars than like the Galilean satellites. It is roughly football-shaped, twice as long as it is wide. The longitudinal axis is about 265 km long and pointed at Jupiter.

By late afternoon, narrow-angle (close-up) images of Io were being displayed on the monitors. People clustered around the TV sets in Von Karman, oohing and ahhing until a sort of sensory overload set in. Reporters and scientists alike seemed to wander around aimlessly with fixed half-smiles on their faces.

That evening, Caltech sponsored a symposium entitled "Jupiter and the Mind of Man." It was a followup to "Mars and the Mind of Man," a 1971 symposium on the eve of Mariner 9's arrival at the Red Planet, and it featured the same cast: Carl Sagan, JPL Director Bruce Murray, *New York Times* science writer Walter Sullivan, science-fiction writer Ray Bradbury, and (by telephone from Sri Lanka) Arthur C. Clarke. Sagan described the evening as "our last moment of innocence" about Jupiter and its family of moons. Murray, looking ahead, commented, "I suspect manned exploration of Jupiter will remain a dubious prospect for a very long time." Due to the intense radiation around Jupiter, Murray predicted, "it will remain the domain of robots."

Bradbury took a more humanistic perspective and talked about the beneficial effects of space exploration for residents of this planet. Optimistically, he declared, "1984 will not happen because of what is happening tonight." Clarke simply warned the people in Pasadena, "Look out for black monoliths!"

Throughout Sunday evening, people came and went, tried to catch some sleep or filled up on coffee from the bottomless Von Karman percolators. During the Viking mission, the coffee machines had shorted out from overuse, but the technology was in better shape for Voyager. By morning, the desks in the press room would look like a sea of white styrofoam cups.

Al Hibbs and his colleagues were on the closed-circuit "Blue Room" television show almost continuously for the entire night. Scientists commented on the pictures as they came in; as the night wore on, their remarks began to sound like the enthusiastic verbiage of baseball "color" commentators ("That looked like a curve ball to me, Howard. . . . He sure hit that one a mile!"). Detailed scientific analysis was all but impossible under the circumstances, and the scientists more or less abandoned themselves to the excitement of the event.

"Jeez, look at that!" exulted Charlie Kohlhase as an enhanced image of Amalthea appeared on the screen. Everyone did look, but by now Amalthea was old news. Attention was focused on the early morning pass at the ever more bizarre Io.

Celebrities and thrill seekers showed up at Von Karman in droves. At one point, California Governor Jerry Brown and his entourage (including ex-astronaut Rusty Schweikart) walked through the press room. While the thrill seekers clustered around the dignitaries, the serious space reporters snarled at them, warning them to get the hell out of the way of the monitors. Governor Brown gave one of his patented off-the-cuff-Zen-mellow talks and predicted that one day there would be a "JPL in orbit." If so, some wondered, would there be adequate parking for the press?

Perhaps as many as a thousand people had press passes that night, and Von Karman was overflowing. The working reporters staked out their desks, chairs, and typewriters and tried to function normally in the midst of the chaos.

Yet there were long gaps when nothing much was happening, either on Earth or on Jupiter. By 1 AM, a general stupor had set in. People sat quietly, staring at TV screens or blank walls, talking very softly, as if they were afraid to wake those who were trying to sleep. *Science News'* space science editor, Jonathan Eberhart, had a long discussion with JPL Public Information Officer Jurrie van der Woude on the relative merits of various brands of Indonesian cigarettes. Space artist Anne Norcia sat quietly, making Mobius strips and origami paper animals. Jeff Lenorovitz of *Aviation Week and Space Technology* left at about 11 PM to play racquetball, and returned at 2:15 AM looking wide awake. On the other hand, Mitch Waldrop, then with *Chemical Engineering News*, returned from a catnap at 2:30 and reported, "I had just enough sleep to make me feel terrible."

Around 4 AM, Voyager made its closest approach to Jupiter at a range of 780,000 km. The event passed almost without notice because of the built-in time lag from Jupiter. Events were recorded in Earth Received Time, which at that point was some 39 minutes later

"It shouldn't look like this at all!" cried Larry Soderblom. From half a million miles, no obvious impact craters were visible on Io. Scientists began wondering about recent volcanism.

than spacecraft time. Also, nothing in particular actually happened at the moment of closest approach. "From the point of view of the Voyager mission as a whole," said Hibbs, "it's simply an interesting point of celestial mechanics."

The Jupiter pictures were routinely spectacular; people were already jaded by the multicolored clouds and baroque flow patterns. At a minimum resolution of 10 km, the cloud tops didn't look much different than they had from a greater distance. The "convective towers" of clouds some had expected were not visible, but there did appear to be a very sharp boundary surrounding the Great Red Spot. The fine details of the cloud structure were not visible; Jupiter was not going to placidly surrender all its mysteries to this invader from the third planet.

Closest approach to Io took place at 7:47 AM at a range of 22,000 km. The resolution of the best pictures was a little more than a kilometer. The limiting factor on resolution was not distance but blur due to the speed of the spacecraft, which went zooming past Jupiter at about 60,000 miles per hour—as fast as any manmade object had ever gone. The on-board computer was set to compensate for the smearing of the images, and nearly all of the pictures came in bright and sharp.

At 5:41, a narrow-angle frame of the Io "hoofprint" arrived. Larry Soderblom noted the "very complex markings" and suggested that "we may be seeing some underlying structure." Like everyone else, Soderblom was waiting for impact craters to show up. By any

FIRST ENCOUNTER

reasonable model of the solar system, there simply had to be impact craters on Io. At one point, he commented on some "pockmarks that resemble impact craters," but whether these were true impacts or not, he couldn't say.

As the curved limb of the moon came into view, Soderblom pointed out that the smooth curve of the horizon was broken by small lumps and knobs. They looked like mountains, "but they would have to be enormous," on the order of 10 km high, which would make them as big as Olympus Mons, the immense volcano on Mars.

Other features were not so easy to interpret. Soderblom described one object as "a bright thing next to a dark thing." That was about all anyone could say. "The surface complexity," said Soderblom, "is mind-boggling."

The pictures were flooding in now, and the astonishment they produced was universal. In the press room, reporters gaped at the monitors and tried to think of new adjectives. Scientists, meanwhile, tried to think of new explanations for Io—because it was becoming increasingly obvious the old explanations had just gone out the window.

"It's a spectacular place," said Soderblom, "but where are all

Io's surface complexity increased as Voyager 1 got closer. From 50,000 miles (left), mountains and plains were visible. From 18,480 miles (right), there was clear evidence of lava flow patterns. The crater is 30 miles across; smallest visible features are about 400 yards across.

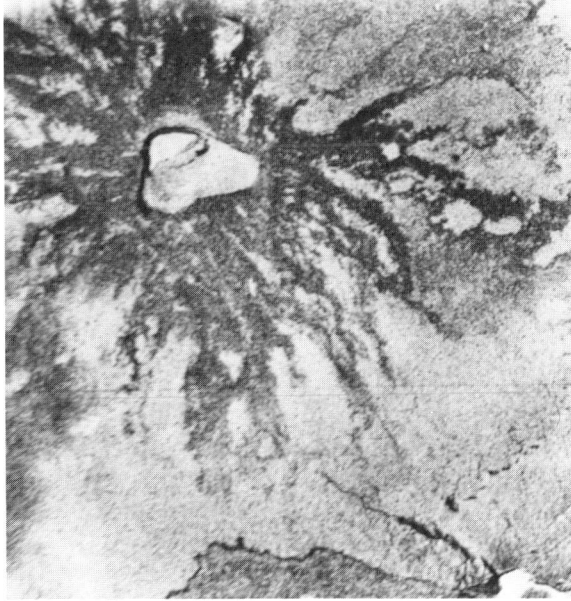

the impact craters? It's clearly a different sort of planet. . . . The absence of craters implies an active planet. . . . These are really incredible!" Io could not have avoided the early cratering episode four billion years ago, but there were no impact craters to be seen. Something had to have obliterated them. Soderblom began to speculate about the possibility of volcanism and rapid erosion.

Soderblom was joined in the Blue Room by Hal Masursky, a geologist who had been counting craters from Mercury to Mars. Io, he admitted, "may be a little more complicated than we suspected." Masursky, Soderblom, and Hibbs watched in wonder as the pictures came in. Their dialogue captures the awe and excitement felt by everyone:

MASURSKY (on seeing a complex flow pattern on Io): "If we didn't know there wasn't water . . ."
SODERBLOM: "No water *now*, apparently, but who knows, earlier?"

MASURSKY: "Larry, are those beginning to look like drainage patterns?"
SODERBLOM: "I thought you'd never say that."
MASURSKY: "This is very important, because there is a bet in the imaging team. Somebody's going to get a bottle of Scotch."
SODERBLOM: "It shouldn't look like this at all!"

SODERBLOM: "Look at that stuff!"
MASURSKY: "We can say categorically: This is not an ancient surface."

And finally, as one more astonishing closeup of the surface of Io arrived, Soderblom shook his head and mumbled weakly, "Oh my God!"

Suddenly, it was over. The incredible, mind-searing encounter with Io was history. Ganymede and Callisto were still to come, but Voyager had already justified its existence. What could possibly top this, everyone wondered?

They would find out in a few days.

Meanwhile, the traditional post-landing/encounter congratulatory press conference was held in Von Karman Auditorium before a packed house of bleary-eyed reporters. NASA and Voyager project officials appeared to say a few words and figuratively pat each other on the back for a job well done. The most quotable lines (as was usually the case at these affairs) came from Noel Hinners, NASA's associate administrator for space science. "Just watching the data come in has been fantastic," he said. "I had a fear that things on the satellites were going to look like the lunar highlands. Nature wins again."

Project Manager Robert Parks then reported that the Jovian radiation was beginning to have an effect on the spacecraft. With about 80 percent of the total dosage already absorbed, Voyager was suffering from an apparent slowdown in a computer timing mechanism. The delay at this point was about six seconds, but within a day the lag had grown to forty seconds. With the spacecraft's two computers out of synchronization with each other, a number of images were taken at the wrong time, some while the scan platform was still moving. Several high-resolution post-close-encounter pictures of Io and Ganymede would be lost because of the problem.

Next, it was the scientists' turn. At the regular 11 AM press briefing, Brad Smith described the feelings of the imaging team: "We're all recovering from what I would call the most exciting, the most fascinating, and what may ultimately prove to be the most scientifically rewarding mission in the unmanned space program. The Io pictures this morning were truly spectacular and the atmosphere up in the imaging area was punctuated by whoops of joy or amazement or both." Smith then showed a full-disk color picture of Io, which he described as "better-looking than a lot of pizzas I've seen."

Larry Soderblom was still fascinated by the lack of impact craters on Io. Unless the rate of cratering in the region of Jupiter was much lower than in the inner solar system—a possibility which completely contradicted existing models—the surface of Io had to be extremely young. Masursky estimated that it could be no older than about 100 million years. "Immense erosion is going on," said Soderblom. The surface gives "a feeling or appearance of corrosion. The surface is generally being gobbled away." Perhaps Io's weird electromagnetic interactions with Jupiter and the torus were responsible —and perhaps not.

The particles and fields men were also excited by the encounter. It appeared that the spacecraft missed the electric flux tube connecting Jupiter and Io, which seemed to be twisting and shifting its

On Ganymede, impact craters at last! Bright rays from the crater at right consist of fresh ice thrown out by relatively recent impacts.

position. Nevertheless, the physicists were confident that they had acquired good data, and they were looking forward to an opportunity to sit down and quietly analyze the information.

That afternoon, the Ganymede pictures began coming in. The closest approach to Ganymede would come at 6:53 PM at a range of 115,000 km. The computer timing problem was apparent in some of the pictures, which caught the moon in the upper-right corner of the frame instead of in the center.

After Io, the geologists and astronomers were almost desperate to see some impact craters. Long-range images of Callisto had suggested the presence of impact craters, but the lesson of Io's youthful surface was fresh in the mind of everyone. There was no substitute for close-up, high-resolution images of a planetary surface.

At last, Dave Morrison spotted an indisputable impact crater on Ganymede. "This is a real find," he said. "We found something on these two bodies we can actually recognize." The total number of craters visible still seemed low, however, and as the spacecraft got closer to Ganymede it became obvious that cratering was only one of the geological processes at work on the surface of the satellite.

"While I'm delighted to see the craters," Morrison admitted, "it's just the opposite of what I would have expected. I was telling everyone a few days ago that I thought Io would have plenty of craters and Ganymede, because of the ice surface, simply would not be able to hold large craters over geological time. So this is fascinat-

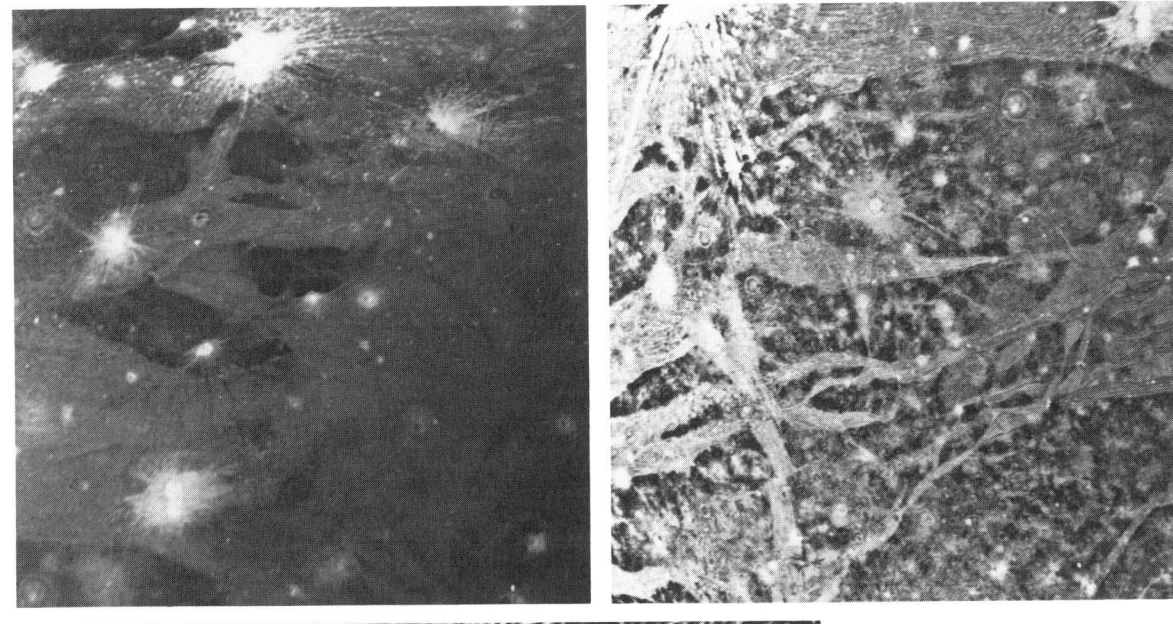

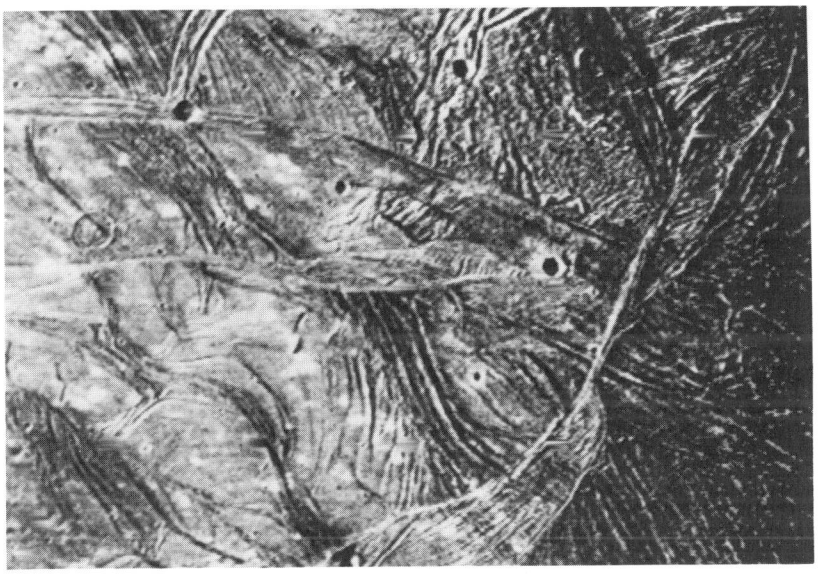

"Dune buggy tracks," faults, ridges, rays— Ganymede, at close range, proved to be a geologist's delight.

ing and this is confusing—both what has happened on Io to erase the craters and why Ganymede's surface is strong enough to preserve them."

The Ganymede closeups were, in their own way, as bizarre as anything seen on Io. The surface seemed to be composed of three types of material. There were large, vaguely polygonal areas of dark (and possibly older) material, broad regions of lighter (and younger?) material, and bright icy streaks and crater rays thrown out by impacts. What was astonishing was the way it all fit together. It looked as if there were fantastic systems of faults and fractures all over the surface of Ganymede, with huge chunks of real estate displaced by many kilometers. It was another first: large-scale tectonic activity had never been observed anywhere but on Earth, where our continents have been sliding around on huge crustal plates for billions of years.

There were also areas of what seemed to be corduroy ridge systems running in parallel lines for hundreds of kilometers. Almost immediately, people began calling them dune-buggy tracks; but there was a lesson in this casual naming of names. In Pasadena, California, dune-buggy tracks seemed like a reasonable analogy; but free-lance lecturer and journalist Jim Loudon looked at the same pictures and saw, not dune-buggy tracks, but snowmobile tracks. Loudon is from Michigan, where snowmobiles outnumber dune buggies. Given the icy surface of Ganymede, Loudon's analogy may have been better; but the main point is that what one gets out of a picture depends to a great

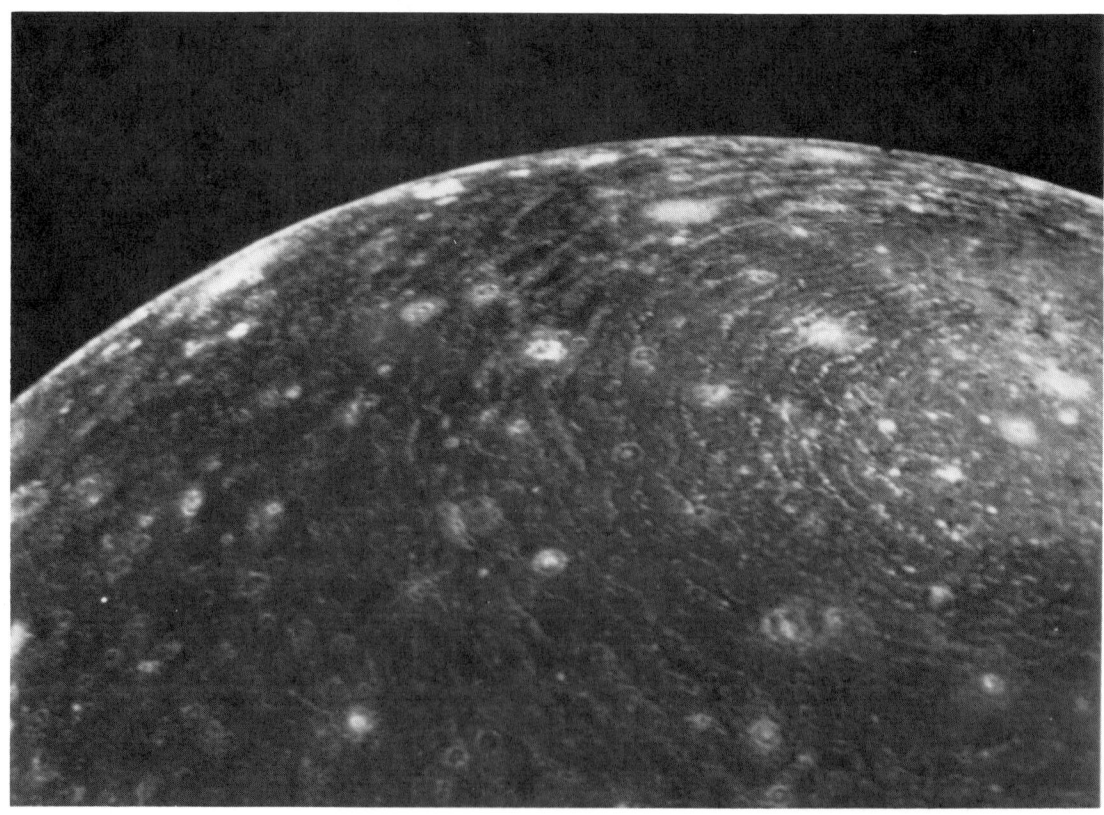

The bull's-eye pattern on Callisto may be the result of a huge, ancient impact. Note spreading "ripples" radiating from the center of the bull's-eye. Bright spots are fresh ice.

FIRST ENCOUNTER

extent upon what one brings to a picture. California chauvinism and Michigan chauvinism are inconsequential; what matters is the preconceptions created by Earth chauvinism.

After barely two days of high-resolution imaging of the Galilean worlds, it was already clear that the scientists' Earth-oriented ideas about what planets are and how they behave would require some major revisions. Because of the great differences in sizes and locations, simple comparisons between Earth and the Jovian moons were likely to be misleading. The solar system was turning out to be far more diverse and complicated than the early missions to the inner planets had suggested.

The next morning, Tuesday, March 6, Voyager 1 passed Callisto at 9:50 AM at a range of 126,000 km. Here, at last, were impact craters—thousands of them. The moon looked as if it had been blasted with buckshot at close range. There was also one huge bull's-eye pattern which seemed to be evidence of an enormous impact. Bright concentric rings, probably of ice, surrounded the bull's-eye, which was about 300 km in diameter.

At the daily press briefing, Larry Soderblom showed the best of the Callisto pictures and declared, "There ain't no such thing as a boring Galilean satellite!"* Callisto had "lots and lots and lots" of impact craters ranging up to 200 km in diameter. What was puzzling —there were *always* puzzles—was the lack of larger craters, excepting the bull's-eye. Judging from Callisto, the cratering rate in the outer solar system was certainly not lower than in the inner solar system, but the *type* of cratering was strangely different. There also seemed to be a complete absence of vertical relief on Callisto. Apparently the icy surface could not support any relief, and whatever was thrown up by the impacts simply sank back down to the mean surface level.

Soderblom also showed color closeups of Io so intensely orange that they almost hurt the eyes. "We're starting to wonder if we should use some sort of color suppression scheme to make them easier to

---

* The press corrected Soderblom's grammar, and the remark has gone down in history as "There is no such thing as a boring Galilean satellite." Jim Loudon showed up at the Voyager 2 encounter wearing a tee shirt bearing the corrected quote, thereby immortalizing it. But, for the sake of historical accuracy, let it be recorded here that Soderblom actually said "ain't."

DISTANT ENCOUNTERS

interpret," Soderblom joked. As for Ganymede, Soderblom confessed that there was still "some confusion as to what is going on." The faulting and fracturing were difficult to interpret, as was the absence of "shoulder-to-shoulder" giant impact craters. Ganymede's surface seemed to have been shaped by processes that had not occurred on either Io or Callisto.

Brad Smith diverted attention from the satellites with some more color images of Jupiter, including some radically stretched false color images that were artistic masterpieces. "If Jupiter had ever posed for Monet," said Smith, "it would have looked like this." Other observers thought they saw Van Gogh's brush strokes in the Jovian clouds.

Smith also presented a color picture of Amalthea, which turned out to be dark red—the darkest, reddest object in the solar system, according to one scientist. Amalthea is also lumpy, irregular, and cratered. It looked like a larger version of Phobos or Diemos, the moons of Mars.

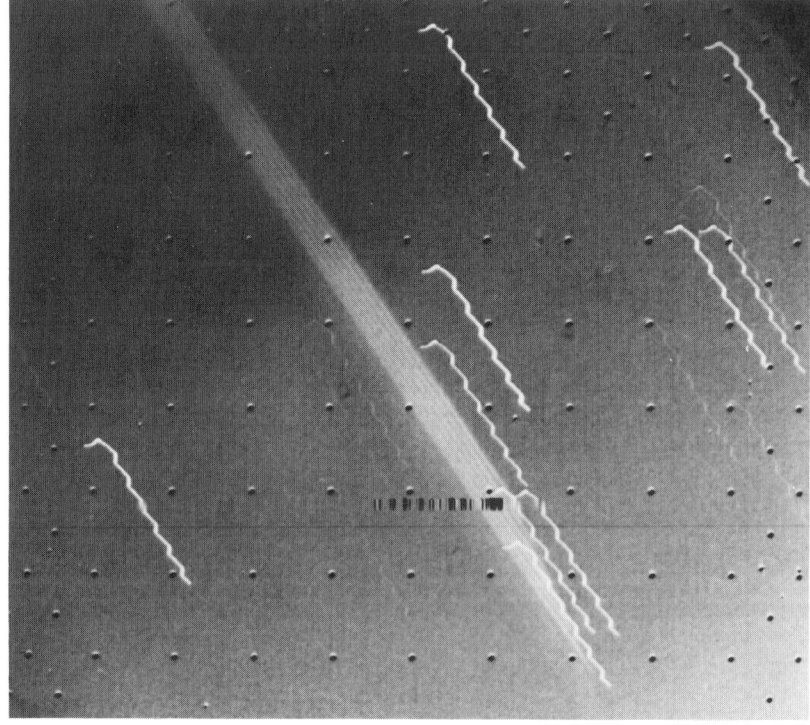

Jupiter's ring was discovered in this multiple exposure. The ring is the faint band in the center of the photograph. Hairpinlike streaks are background stars.

112

The particles and fields men were also heard from. Lyle Broadfoot reported extremely intense auroral emissions in the Jovian polar regions. The emissions seemed to come from excited molecular hydrogen.

IRIS Team Leader Rudy Hanel announced infrared readings that indicated that the temperature over the Great Red Spot was consistently 3°C colder than elsewhere, implying that the Spot was higher than the surrounding atmosphere. He also reported that there seemed to be some "hot spots" on the surface of Io. The significance of that observation would be appreciated a few days later.

The next day, March 7, Brad Smith made the formal announcement of the discovery of the Jovian ring. It came as no surprise to many members of the press corps, who had heard of the discovery shortly after it happened on Sunday, the 4th. But the announcement had been delayed so that the scientists could be absolutely certain that they had truly found a ring.

The eleven-minute time exposure showed six separate images of the ring, which looked like nothing more than a bright streak. There were also hairpinlike trails of stars in the background. The spacecraft had been "nodding" during the exposure, resulting in the multiple images of ring and stars. The nodding was fortuitous, because without it the ring might have been difficult to see as a single exposure.

The discovery was of fundamental importance, rather as if a photo from an Earth satellite unexpectedly revealed a major new land mass just west of California. For centuries, Saturn was presumed to be the only planet with a ring system. Theories were evolved which endeavored to explain why Saturn was the only planet that *could* have rings. Then, in 1977, astronomers in an airborne observatory over the Indian Ocean detected a set of rings around Uranus. Now, there was one more ring system to be explained. Neptune was the only remaining gas giant not known to have rings, but it seemed that it was just a matter of time before someone (or perhaps, Voyager 2 in 1989) found rings there, too.

Perversely, all three ring systems were unique, none resembling the others. Saturn's rings were broad and spectacular. The rings of Uranus consisted of nine dark, narrow, concentric rings arrayed around the planet in a bull's-eye pattern. Jupiter's ring looked more like a single-strand necklace, thin and ribbonlike. It could be no more than about 30 km thick, and judging from the way it scattered light, was apparently composed of extremely small particles.

Smith admitted that "we were very lucky" to find the ring. "Incredibly, the edge of the ring fell in our field of view." The long-exposure picture had been planned with no real expectation that it would capture a ring or anything else.

Ironically, the discovery made obsolete the diagram of the solar system contained in the Voyager record. The Pioneer plaques had depicted only the Saturnian rings. The Voyager record updated the plaques with the Uranian rings. Now it was beginning to look as if *all* gas giants might have rings. Earth-based searches for Neptunian rings had failed to find them, but Jupiter's ring hadn't been seen from the ground, either.

The discovery of the ring showed the value of sending two spacecraft to Jupiter. Now that they knew it existed and where it was, the scientists could reprogram Voyager 2 for a sequence of more detailed ring images. By the end of the week, in fact, two astronomers at the University of Hawaii's Mauna Kea Observatory managed to detect the ring from the ground.

On March 8, the final science briefing was held. Thirteen scientists trooped up to the Von Karman stage in two shifts and presented as much information as they could in a limited time. Andy Ingersoll summed up the general situation for everyone: "We are buried under a mountain of data." It would take months or years to sift through that mountain, and in the meantime, the experimenters could do no more than present "instant science." The situation was not entirely satisfactory for either the quote-hungry press corps or the naturally cautious scientists. The Voyager 1 Jupiter encounter had been a profound experience for all concerned, and what was needed now, more than anything, was time for quiet reflection on the meaning of it all.

"We have had almost a decade's worth of new discoveries in the last two weeks," said Project Scientist Ed Stone. "I think that all the people who have been talking to you feel the same saturation of new information." The press felt it, too, for that matter, and it was a weary, slightly dazed band of journalists that packed up their typewriters and tape recorders that afternoon. The encounter was over—it seemed.

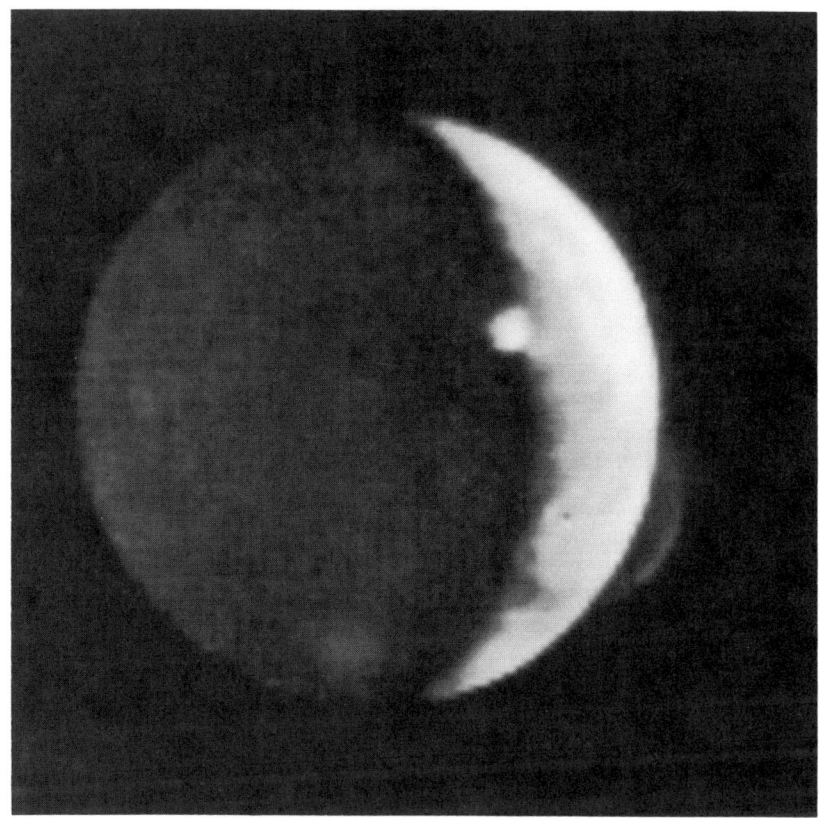

Linda Morabito discovered Io's active volcanism in this dramatic image. Two simultaneous eruptions are visible, at lower right and along the day–night terminator. Six more active volcanoes were later found in Voyager 1 images.

As the press was packing, a JPL member of the optical navigation team, Linda Morabito, began to work with a newly acquired image of Io. It was a long-range time exposure, designed to show the stars around Io. Once the stars were identified, their precisely known coordinates could be used to plot the exact position of the spacecraft when it took the image. This was a matter of some importance, since Jupiter's gravitational field had already flung the spacecraft toward its 1980 rendezvous with Saturn; arrival time at the sixth planet would depend on the time and place of departure from the fifth.

Working at her computer console, Morabito played with the controls to bring out maximum detail in the picture. To her surprise, she noticed a crescent-shaped blob hugging the limb of Io. The blob was huge, more than a hundred kilometers high—that is, if one interpreted it as a cloud on Io. Airless Io ought not to have clouds.

The next day, Friday, Morabito and her team colleagues examined the picture more closely and tried to eliminate other possibilities, such as computer error or smearing of the image. One by one, the other explanations were excluded, leaving only the obvious answer: Morabito's cloud was actually a huge volcanic plume spurting from an active volcano on the surface of Io.

Most of the imaging team had already gone home for the weekend, so it was not until Sunday night that Brad Smith saw the picture. By early Monday morning, imaging team scientists were poring over other Io pictures, looking for more evidence of active volcanism. They quickly found it. By week's end, eight active volcanoes had been identified, one of which was in the center of the original "hoofprint". The news was released to the public on March 12, much to the consternation of journalists who had just left JPL.

Now that scientists had seen the incredibly young surface, the existence of volcanoes on Io was not surprising. Indeed, volcanoes were about the only plausible explanation for the total absence of impact craters. Volcanoes were also plausible on purely theoretical grounds. The March 2 issue of *Science* had contained a remarkably well-timed paper by Stanton Peale, Pat Cassen, and Ray Reynolds, in which the authors argued that Io might be caught in the middle of a gravitational tug-of-war between Jupiter and the other Galilean satellites, particularly Europa. The friction produced by the opposing forces could have melted Io's interior, producing "widespread and recurrent surface volcanism." The paper, written months earlier, suddenly seemed to be a prophecy come true.

But absolutely no one had expected to see Voyager images of an active eruption—let alone, *eight*. On Earth, there might not be eight major eruptions in an entire century. When Mount St. Helens exploded the following year, it was a front-page story for days. There are plenty of volcanoes on Earth, but at any given moment it is unlikely that even one of them will be active. Yet this small moon of Jupiter was sporting eight simultaneous eruptions.

Confirmation, if any was needed, came from John Pearl of the IRIS team. Thinking about the infrared hot spots detected on Io, Pearl had come to the conclusion that they must be caused by active volcanism. When the IRIS results were compared with the Io images, there was a close match between the hot spots and the volcanic plumes.

The Io volcanoes would provide answers to many of the questions raised during the encounter, but the volcanoes themselves posed many more questions. How did the volcanoes operate? What was the chemical nature of the lava? How quickly could the eruptions obliterate impact craters? The media show, the fireworks, and the mind-numbing revelations of the Voyager 1 Jupiter encounter were over now. It was time for the scientists to return to their laboratories and computers and try to make sense of what had just been learned. After years of sowing, cultivating, and weeding in their scientific gardens, the harvest was in, and it was the greatest bumper crop in history. Jupiter was a feast for the mind, a feast for the senses.

And this was just the first course.

6 ATTACK OF THE SPACE GYPSIES

A headline clipped from the *National Enquirer* hangs on the wall above the desk where this book is being written: "Top American Space Scientists Discover . . . INVISIBLE ALIENS FROM SPACE LIVE AMONG US."

Well, probably not. But top American space scientists *do* worry about this sort of thing—the headline, not the invisible aliens. In an enterprise funded almost entirely by taxpayers' dollars, public perception is critical to the survival of the American planetary exploration program. What the public reads and hears about a spacecraft may be more important than what the spacecraft actually *does*. Scientists are becoming more aware of the importance of the media, but they are not always comfortable with it. The relationship is not unlike that between a Hollywood starlet and the show-biz press. She wants to be taken seriously as an actress, and they just want to run pictures of her legs.

The Voyager press corps is a motley crew of journalists. On a typical day at JPL's Von Karman Auditorium, one would encounter hard-boiled newspapermen straight out of *The Front Page*, dedicated young correspondents for obscure scientific periodicals, science-fiction authors, overbearing television correspondents complete with their camera crews and "go-fers," bewildered feature writers, a small army of Japanese reporters who speak little or no English, and clusters of starry-eyed space enthusiasts who know every line of dialogue in both *Star Wars* movies.

Once a day, usually at about 11 AM, the denizens of the press room grab their notebooks and tape recorders and make their way into the auditorium for the day's press conference. Depending on what is happening, there may be anywhere from fifty to five hundred people waiting to hear the words of four or five scientists. The first event is the official briefing. The throng slowly settles in the folding chairs as a NASA or JPL public information officer goes up to the podium at the left of the stage, taps the microphone a few times, then introduces the first speaker.

Almost without exception, the first speaker is the project manager or one of his deputies, reporting on the health of the spacecraft. Unless there is a specific glitch to discuss, the spacecraft report is brief and boring, and the reporters simply scribble "All is well" in their notebooks.

The first scientist introduced is usually a member of the imaging team, most likely Brad Smith. Smith makes a few general comments, throws out a sprinkling of carefully crafted quotable lines, then asks for the first picture. The curtain is drawn, revealing a large movie screen, and the lights go down, making it almost impossible for the reporters to take notes. The first picture is projected on the screen. Smith takes a flashlight and points out interesting aspects of the image. The first few images are interesting, but Smith usually saves the best for last. The reporters see the raw images displayed as they come in from the spacecraft, but they seldom see the computer-processed color images until the press conferences. When Smith shows his prize picture of the day, there are oohs and ahhs from the press, and sometimes applause.

Smith may be followed by another imaging team scientist, usually Larry Soderblom, with more images to show and discuss. Generally, Brad Smith talks about the planets and their rings, while

Soderblom focuses on the satellites. Following the imaging team, two or three particles and fields men report on their latest findings. They seldom have pictures to show, but they are never without charts and viewgraphs to illustrate their data. Frequently, the physicists spend five minutes describing the workings of their particle detection instruments and only two minutes reporting their data. At other times, the press may be subjected to a long, densely detailed analysis of the meaning of a particular bit of information. As a rule, the particles and fields men are older than the astronomers and geologists of the imaging team and seem less comfortable in front of the press, less sensitive to the needs of this particular audience. It is also true that few members of the press corps have a strong enough scientific background to be able to appreciate and understand the subtleties of magnetospheric physics.

When the formal presentations have ended, the scientists return to the stage and sit behind a long desk, with nameplates and microphones, while the public information officer or PIO calls on the reporters for questions. The press conferences are videotaped and sometimes broadcast live to reporters at NASA facilities on the East Coast. Especially important briefings may be carried live on public television. JPL workers pass cordless microphones to those with questions, and the logistics of getting the mikes to the right people frequently produce delays and confusion. The PIO in charge, usually Frank Bristow of JPL, stands at the podium and selects the recipients of the microphones.

The first question usually comes from Jonathan Eberhart of *Science News*. Eberhart, in his late thirties, has been covering space since the early sixties and is undoubtedly the most knowledgeable member of the press corps. He wears his hair in a long, gray pony tail, has a full beard, and looks more like a folksinger than a science reporter. In fact, Eberhart *is* a folksinger, and when he comes to LA for a mission he usually manages to squeeze in a couple of performing gigs. His "old songs for tomorrow," a collection of space folk songs, are well known at JPL and he is often asked to perform them. During the Voyager Saturn encounters, Eberhart and some musically inclined journalists and JPLers even organized their own rock band, inspired by and named after the only distinctive feature seen on Saturn's largest moon—the Titan Equatorial Band.

Eberhart is usually followed by the reporters from the big daily newspapers. Reporters from Eastern afternoon papers are already facing deadlines by this time, and they occasionally have to leave the briefing to file their stories. The regulars include George Alexander (*Los Angeles Times*), Tom O'Toole (*Washington Post*), John Noble Wilford (*New York Times*), Bob Cooke (*Boston Globe*), Dave Perlman (*San Francisco Chronicle*), Richard Saltus (*San Francisco Examiner*), and Dave Salisbury (*Christian Science Monitor*). Other reporters representing smaller dailies are usually present, as are the frenetic wire service correspondents, who always seem to have something to file.

After calling on the newspapermen, Bristow usually gives the nod to the magazine correspondents, such as J. Kelly Beatty (*Sky and Telescope*), Mitch Waldrop (*Science*), Richard Berry (*Astronomy*), and Dennis Overbye (*Discover*). There are also numerous free-lancers who have come to JPL with press credentials gleaned from local newspapers, Sunday supplements, and FM radio stations. The free-lancers, who seldom have expense accounts or travel vouchers, get to JPL any way they can because they want to *be* there when the solar system is unveiled.

The Q & A session continues for perhaps twenty minutes before Bristow brings the proceedings to a close. Typically, the entire press conference lasts from one to two hours, depending on how much new information there is to present. During the Q & A, there are questions from twelve to fifteen reporters; most of the journalists simply take notes and save their questions for one-on-one interviews.

After the briefing ends, about a dozen reporters gather at the stage, asking followup questions of the scientists, who sometimes find themselves trapped on stage for fifteen minutes or more. If a scientist manages to elude the detailed technical questions of the more scientifically minded journalists, he may still be snagged by a television reporter as he tries to exit from the auditorium. The TV crews film standup interviews with their subjects posed in front of the full-scale Voyager mockup suspended from the Von Karman ceiling.

Standing in the no-man's land between the reporters and the scientists are the public information officers of JPL and NASA. Their efforts during the Voyager mission were nothing short of heroic. Keeping hundreds of journalists supplied with up-to-the-minute information is, in itself, no easy task. But the PIOs must cope with the diverse demands of newspaper, magazine, book, wire service, radio, local television, network television, foreign television, and unclassifiable free-lance journalists, all of whom have different needs and deadlines. They must also assure the availability of scientists, whose schedules may be even more chaotic than those of the press. And, most important, they keep the Von Karman coffee machines running.

As the journalists make their way out of the crowded auditorium, PIO Jurrie van der Woude gets ready for the onslaught at the picture desk. On a typical day, a packet of about six pictures will be released to the press, and there is almost always a mad scramble for the packets. The Image Processing Laboratory works overtime during an encounter to produce "hard copies" for the press and for NASA officials in Washington. It sometimes happens that the scientists themselves can't get the photos they want because the IPL is too busy preparing the press package.

Reporters who have to file stories usually return to their typewriters following the briefing. JPL provides typewriters for the press, but many bring their own typewriters, and an ever-increasing number of journalists are equipped with portable computer terminals.

For lunch, many of the reporters go to the JPL cafeteria where the food is surprisingly good. In the cafeteria, the journalists share tables with scientists and JPL employees. It frequently happens that a reporter learns more at lunch than at the press conference. Away from the lights and cameras, the scientists relax and become more talkative about their work.

Back in Von Karman, afternoons are a time for interviews, bull sessions, and occasionally, writing. People wander around the press area, drinking coffee, comparing notes, and staring at the monitors. The Public Information Office provides the journalists with minute-by-minute mission timelines, so at any given moment the reporters know precisely what the spacecraft is doing and what picture sequence is being received. Journalists with nothing better to do hang around simply to watch the pictures come in.

The Jet Propulsion Laboratory is located at the head of a dry arroyo just northwest of Pasadena, in the foothills of the San Gabriel Mountains. It is really a small city in itself, and the work conducted there is not limited to planetary exploration. JPL is one of the centers of solar energy research, has an advanced robotics laboratory, and is a mecca for computer graphics experts. There is always something interesting going on there, and many reporters who come for a planetary encounter may leave with two or three additional stories about completely unrelated topics.

Evening finds the journalists embarking on mass expeditions to local restaurants and bars. On nearby Foothill Boulevard, one can find Mexican, Thai, Chinese, Japanese, and American fast-food restaurants; dieting space gypsies have a tough time of it during encounters. There are also several parties thrown during encounters, adding to the general wear and tear on minds and bodies.

When a journalist is finally ready to retire for the night, or what's left of it, his bed may consist of a borrowed couch or a sleeping bag on someone's floor. Reporters for major magazines and newspapers usually stay at the Hilton or the Holiday Inn in Pasadena, but many members of the press corps have limited budgets and no expense accounts. They drive to LA and stay with friends, share cheap motels, or sometimes even camp out. This low-rent gypsy lifestyle has its drawbacks; one reporter was attacked and seriously beaten while staying in a cheap motel near the Rose Bowl.

Coming to JPL for a planetary encounter is an intense, highly charged experience—two weeks of life in the fast lane for journalists and scientists alike. For veterans, encounters provide opportunities to renew old friendships and recapture the excitement of previous missions; for newcomers, the scene at JPL can be overwhelming.

Veteran science writer Dennis Overbye first joined the JPL space gypsies during the Voyager Saturn encounters, and soon pronounced himself "hooked"; he had joined the ranks of what Jonathan Eberhart calls "the space junkies." Like dozens before him, Overbye discovered that nothing else in life is remotely comparable to a planetary encounter at JPL. At one party during the Voyager 2 Saturn encounter, as the Titan Equatorial Band belted out rock 'n' roll and journalists and scientists mingled freely, Overbye turned to another reporter and said wistfully, "You know, it's times like this when I think about the people with 'straight' jobs."

Whatever the fringe benefits, journalists come to JPL to do a job. That job is to report and explain the discoveries of the planetary scientists and their spacecraft. But the scientists are also there to do a job, and for many of them the needs of the media are an unwelcome distraction. Unraveling the newly revealed mysteries of the universe is a difficult task under any circumstances; to do it on a lighted stage before the eyes of millions is all but impossible.

During the Viking mission to Mars, Norman Horowitz, a biology experimenter, expressed the scientists' difficulties. "If this were normal science," he told a press conference, "we wouldn't even be here—we'd be working in our laboratories for three more months. You wouldn't even know what was going on, and at the end of that time we would come out and tell you the answer. Having to work in a fishbowl like this is an experience that none of us is used to. . . . You're looking over the shoulder of a group of people who are trying to work in a normal way in an abnormal environment."

Scientists, like other human beings, do not enjoy being wrong in public. In the days before a planetary encounter, a scientist is likely to be torn by conflicting emotions. On one hand, he is filled with the excitement of anticipation, knowing that a subject he has spent years studying is about to be transformed in some wonderful, unpredictable fashion; on the other hand, he is aware that his laboriously constructed theories are very likely to be blown out of the water. And through it all, reporters will be thrusting microphones in his face and demanding answers to questions no one had ever thought to ask until now. He must cope simultaneously with detailed, sophisticated probing by experienced science reporters as well as scientifically naive questions from less knowledgable journalists. He must explain complex scientific matters without sounding like an elitist egghead and describe mind-boggling new phenomena without coming across as an excitable schoolboy. He wants to be right, yet he knows he will be wrong.

Being wrong is an accepted part of the business of science. The scientists understand this and realize that it is no disgrace to have predicted impact craters on Io or not to have predicted a ring around Jupiter. In the long run, scientists probably learn more from being wrong than from being right.

The scientists, however, worry that the journalists will be insensitive to this point—as some certainly are. "INVISIBLE ALIENS FROM SPACE" is an extreme example, but even journalists for more respectable publications sometimes distort or misinterpret the words of the scientists. Informing the public about scientific *results* is not enough; the underlying scientific philosophy and method are equally important.

Some Voyager scientists felt that they had been "burned" by telling the press too much too soon in previous missions. For example, there was the "moon of Mercury" controversy during the Mariner 10 mission. For a brief time, it looked as if the spacecraft had discovered a small Mercurian moon; later, the moon turned out to be a star that only seemed to have moved as a satellite would, due to some odd geometry between the spacecraft and the planet. It was a relatively minor matter, but it was embarrassing to have to retract news of the "discovery." Fallout from such episodes could be seen during the Voyager 1 Jupiter encounter. The formal announcement of the discovery of the Jovian ring was delayed for several days while scientists made absolutely certain of what they were seeing. When Brad Smith finally delivered the news, he did so with all the verve and joy of a Commerce Department official announcing another increase in the cost of living. Many reporters felt it was an odd performance, even for a scientist.

In fact, relations between the scientists and the press seemed somewhat strained throughout the first Voyager encounter. Some reporters complained that they were being treated like subjects in a sensory deprivation experiment. Before the encounter, the scientists —understandably—were reluctant to speculate; afterward, many of them seemed frustratingly tight-lipped about the meaning of it all. In response, several journalists organized informal "backgrounders" in which a group of scientists would meet with fifteen or twenty reporters and discuss a general topic such as satellites or atmospheres. In this more relaxed format, the scientists and reporters got to know each other better and communication was improved.

As the Voyager mission moved on from Jupiter to Saturn, relations between the press and the scientists evolved and improved. People knew each other better and had more realistic expectations. Also, there was more of a feeling that this was a shared enterprise, that everyone was in the same boat. Scientists were less worried about seeming to be wrong, and the reporters came to realize that for many of their questions answers simply didn't exist.

Both the scientists and the veteran reporters felt that the expertise of the press corps had improved substantially since the days of the Viking mission. The sudden proliferation of popular science magazines helped, too. This new evidence of increased public interest injected fresh life into the JPL press corps. In addition, many of the Voyager scientists found themselves writing articles and books for general, nonscientific audiences, and the experience seemed to give them a better appreciation of the task faced by the full-time journalists.

In the interactions between scientists and journalists, much depends on individual personalities. During the long Voyager mission, Brad Smith, as head of the imaging team, emerged as the star of the show, if only because he was usually the man behind the microphone when some bizarre new discovery was announced. To many reporters, Smith seemed a bit crusty and unapproachable, but following the Jovian ring episode, he seemed to loosen up considerably. By the time of the Saturn encounters, he looked like a man who was secretly enjoying himself but would be damned if he'd admit it. Other scientists projected their own personal quirks and traits. Larry Soderblom was revealed as a man with a fine sense of the absurd; David Morrison was a source of both articulate, well-thought-out quotes and boyish enthusiasm; atmospheric scientist Andy Ingersoll explicated the mysteries of gas giants with a sharp, dry, Doonesbury-like wit; and Project Scientist Ed Stone cruised through the confusion with balance, grace, style, and a seemingly endless supply of patience.

The press corps also had its share of distinctive personalities; ABC-TV's Jules Bergman, long known as the Howard Cosell of science journalism, may have been the most distinctive. Bergman's relationship with the scientists was stormy, and at one point a scientist publicly chided him for not doing his "homework." But as the mission continued, Bergman caught up on his homework and, during the Voyager 2 Saturn encounter, enjoyed a brief moment of triumph when NBC-TV's Roy Neal asked one of the unanswerable "what does it all mean?" questions for which Bergman was famous. When the microphone was next passed to Bergman, he began, "Now that NBC has asked a 'Jules Bergman' question . . ." The ensuing laughter from the press corps helped to soften Bergman's reputation; the man could laugh at himself.

Bergman, of course, was the personification of the electronic media—highly visible, highly paid, and often highly resented by the print journalists. The demands of radio and television journalism frequently force the electronic reporters to ask sweeping, Bergmanesque questions. These "film at eleven" questions are often annoying for both the scientists, who get weary of explaining what it all means for the man in the street, and the print journalists, who have a greater need for "hard" information. But the general consensus seems to be that the electronic media's coverage of science has improved substantially in recent years, at least partly due to the excitement created by Voyager. The spectacular images of Jupiter and Saturn made Voyager a uniquely attractive television event, and the networks responded with expanded and relatively detailed coverage.

For the most part, print and electronic journalists co-existed peacefully, like different species sharing a waterhole. They were all there, one way or another, to Get the Story. And the story itself could be viewed from any number of angles, whether by correspondents for science magazines or by *Eyewitness News* crews. Perhaps the only common element was the universal desire to get an interview with Carl Sagan.

By the time of the Voyager encounters, Carl Sagan had emerged as a full-blown Personality. Through his best-selling books, his highly visible role in the Viking mission, and—especially—his regular appearances on the *Tonight Show*, Sagan had become the most famous scientist in America. Virtually every journalist who came to JPL hoped to glean a few Saganisms.

But Sagan, although he was a member of the imaging team, played a relatively minor role during the Voyager Jupiter encounters. Most of his time and energy were absorbed by the creation of *Cosmos*, his thirteen-part series for public television. Filming for the series began as Voyager 1 was approaching Jupiter, and continued for nearly a year. By the time Voyager 1 reached Saturn, in the fall of 1980, *Cosmos* was on the air.

The idea of *Cosmos* was irresistible. At one time or another, virtually everyone in the space business—scientists and journalists alike—had anguished over the lack of quality science programming on television. If someone cared enough to do it right, television could be a spectacularly effective medium for the presentation of science. And suddenly, here was Carl Sagan, with an $8.5 million budget and thirteen hours of prime time to give life to what was, whether he realized it or not, a nebulous group fantasy. Human nature being what it is, the *Cosmos* project inspired equal amounts of admiration, envy, and resentment.

Perhaps inevitably, Sagan and his television series became the target for a good deal of snide sniping by journalists and other scientists. Sagan had carved out an utterly new piece of turf for himself, and was standing there all alone. Was he a scientist, a television performer, or as one critic put it, an "astronomer rock star"? In the fall of 1980, Sagan appeared on the cover of *Time*, wading along the shore of the "cosmic sea"—actually the surf off Santa Monica. The magazine appeared while a conference of planetary scientists was in progress in Tucson. One scientist rushed around showing the cover to everyone, saying, "See? He *can't* walk on water!"

But there is a revealing sidelight to this anecdote. The scientist had bought his copy of *Time* at the small newsstand in the Ramada Inn where the conference was being held. All the remaining copies of the magazine quickly disappeared, all except one. The elderly woman who tended the newsstand wanted to save the last copy for herself. She had heard Sagan would be attending the conference and she was determined to get his autograph on the *Time* cover. "He's the only reason I ever stay up for the *Tonight Show*," she said. "I think he's great."

Like the woman in Tucson, millions of Americans had become Carl Sagan fans. To a public that had grown accustomed to seeing scientists only as apologists for disasters like Three Mile Island or as naive or maniacal characters in Hollywood thrillers, Sagan was a unique and refreshing personality. Here was an actual working scientist who could laugh and joke and speak comprehensibly. He was unashamed to admit his enthusiasm and his love for science, his "zest" for exploration. For millions whose only previous contact with science was in uninspired high school courses, Sagan actually made science seem exciting.

While some other scientists might grouse that they could be just as stimulating and amusing if someone would only put *them* on television, and while critics carped that *Cosmos* was uneven and too "personal" (the show was subtitled "A Personal Voyage"), Sagan had nevertheless made a significant contribution to science. *Cosmos* was seen by tens of millions of people around the globe, and it was Sagan's hope that at least some of those viewers would be inspired by the series to make science their career. Whether or not that happened, the series exposed people to the workings of the scientific mind and the demands of scientific method. At a time when the scientific illiteracy of Americans is becoming a national scandal, *Cosmos* struck a solid blow against the know-nothings, the mystics, and the pseudo-scientists.

Sagan's scientific colleagues are well aware of the importance of his work. Even his critics recognize that Sagan's powerful advocacy of space exploration has been vital to the continuation of the American planetary program. In 1980, at a time when NASA's planetary budget (just a few percent of NASA's total budget, which is less than one percent of the overall federal budget) was being cut to the bone, Sagan and JPL director Bruce Murray founded the Planetary Society. A year later, the pro-space organization had 100,000 members, making it the fastest-growing organization of any kind in America. To Sagan, this was strong evidence that public support for space exploration was increasing, not declining.

Sagan's public image makes it difficult to focus on his purely scientific achievements, which are substantial. During the hectic Voyager 1 Jupiter encounter, as *Cosmos* was just getting rolling, Sagan somehow found time to co-author a scientific paper for *Nature* on sulfur flows on the surface of Io. He was among the first to successfully explain the atmosphere of Venus and the surface processes at work on Mars. His insights concerning the Saturnian moons Titan and Iapetus have contributed greatly to our post-Voyager understanding of those strange bodies. His concern—obsession might not be too strong a word—with the origin and nature of life has done much to make exobiology a scientifically respectable subject.

It is probably no exaggeration to say that, whatever their reservations, everyone at JPL during the Voyager encounters felt a kinship to Sagan. Better than anyone, Carl Sagan expressed the dreams and aspirations that drew people to JPL . . . and to Jupiter. Without being mawkish or melodramatic, he could evoke the passionate spirit of the explorers and scientists of olden days, and the dedication, excitement, and awe that fueled Voyager. There was, in all of us at the Lab, a sense that we were involved in an enterprise of historical dimensions, and that the exploration of space was a noble, glorious, and quintessentially human activity. Simply and eloquently, Carl Sagan reminded us of what it was all about.

In a sense, Sagan, the other scientists, and the space gypsies were all acting as tour guides for the rest of their fellow passengers on Spaceship Earth. Despite having to face an audience of four billion people, to all those involved the daily clash of egos, the pressure of deadlines, and the grind of life in the fast lane all came to seem no more than trivial distractions. The story was what mattered, and long after all the headlines and bylines have been forgotten, the story will remain.

7 INSTANT SCIENCE

For the fourth time in history, a robot explorer from Earth sailed toward a rendezvous with Jupiter. As Voyager 2 entered the Jovian magnetosphere in the first week of July, 1979, no one was exactly blasé about the planet, though there was a feeling at JPL that this time Jupiter was familiar turf. After the mind-blowing Voyager 1 encounter, it seemed unlikely that Jupiter could come up with anything new to match the discovery of rings and active volcanoes. Voyager 2 would show us the opposite hemispheres of the Galilean satellites and provide the first close look at Europa; but somehow, the entire encounter had the air of a summer rerun.

Much more attention was focused on the spacecraft itself. Voyager 2's balky radio receiver was still causing problems, and how it would react to the Jovian radiation environment was a continuing worry for project personnel. Since Voyager 2 would not be passing as close to Jupiter as did its sister ship, its total radiation dose would be smaller; however, the general condition of the spacecraft was worse, and Voyager 2, destined for encounters with Uranus in 1986 and Neptune in 1989, had much farther to go. The Uranus option was still alive, but a major malfunction at Jupiter could kill it.

The space gypsies at JPL also had other things to think about that summer. Pioneer 11 was now just two months away from mankind's first encounter with Saturn. The Voyager 1 Jupiter encounter had whetted everyone's appetite for exploring the unknown, and Saturn now seemed more enticing than Jupiter.

And then there was Skylab. While Voyager 2 was zooming toward its first planetary encounter, the abandoned space laboratory was slowly sinking toward a fiery final encounter with Earth. Higher than normal solar activity had "inflated" Earth's upper atmosphere, causing increased drag on the spacecraft. Launched in 1973, Skylab had been expected to remain aloft into the eighties; but now it was coming down early, and no one knew where. The triumph of another successful Voyager encounter would be utterly ignored if debris raining down from Skylab should hit a populated area. Although the odds against anyone being hit with a piece of Skylab were overwhelming, people everywhere were looking up nervously and wondering if Chicken Little had been right.

Von Karman Auditorium opened for the press on Thursday, July 5. That afternoon, imaging team member Garry Hunt, wandering around the press room, asked a PIO, "You're not going to let Skylab steal all the attention, are you?" The PIO shook his head and replied, "I don't know. It's looking bad." One of the television monitors in the press room had been hooked up with a direct link to NASA's Skylab Watch, and many reporters were clearly paying more attention to Skylab than to Voyager. Most of the journalists were well aware that Skylab represented an extremely small threat to human life ("What if it hits a whale?" someone asked), but to their editors, the Skylab story took precedence over another Jupiter encounter. Meanwhile, the NASA bureaucracy was collectively trembling over the thought of the potential consequences of a Skylab disaster. *Washington Post* science writer Tom O'Toole joked that if the worst happened, they would send in a "NASA SWAT team—two engineers, a lawyer, and a PR man."

The first press conference of the encounter was held on Friday, July 6. Ray Heacock, the new project manager (he replaced Robert

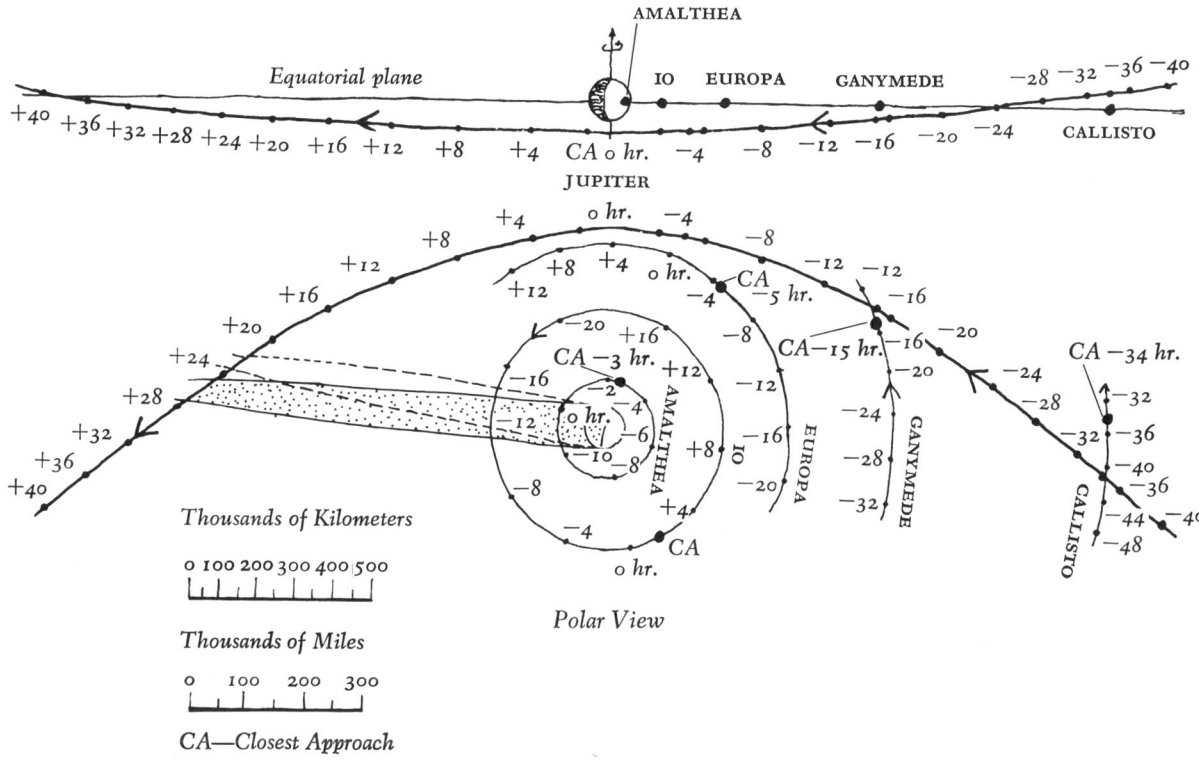

Voyager 2 Flyby of Jupiter July 7–11, 1979

Parks, who had replaced John Casani), said that Voyager 2's radio receiver posed "no serious threat," although he admitted that "we have stretched the system almost to the limit." Heacock and his people had such confidence in the spacecraft and their ability to stay in touch with it that they rescheduled a post-encounter engine burn to just two hours after closest approach to Jupiter. The seventy-six-minute burn would put Voyager 2 on course for Saturn. By conducting the burn so soon after closest approach, they would save some twenty-one pounds of hydrazine fuel, increasing the amount of fuel available for the Uranus option.

It had been apparent for several weeks that Voyager 2 was seeing a different planet than was observed by Voyager 1. The large white ovals observed near the Great Red Spot had drifted around the planet; there was still a large oval south of the spot, but it was a different oval. The turbulence to the east and west of the spot seemed to be breaking up, and the spot itself was a more uniform color. To Brad Smith, it looked as if Jupiter was reverting to its blander Pioneer 10 appearance.

The magnetosphere had also changed. Voyager 1 crossed the

The Voyager 2 trajectory at Jupiter featured a close pass at Europa, which was seen only from a distance by Voyager 1. Note how Jupiter's gravitational field bends the spacecraft's flight path, sending it on toward its next target, Saturn.

From 8 million miles, Voyager 2 saw a somewhat less turbulent Jovian atmosphere than its sister ship. Photogenic Io again intruded.

bowshock five times before entering the magnetosphere to stay at 47 R_J; Voyager 2 crossed it eleven times but entered the magnetosphere at a distance of 62 R_J from the planet; the solar wind was more variable now. The high-speed super-hot sulfur and oxygen ions from the Io torus were less energetic than they had been, although they were still present. Surprisingly, although the magnetosphere was more distended than it had been in March, the radiation levels inside had "hardened." The spacecraft was going to get a larger dose of radiation than anticipated, though it would still get less than Voyager 1.

The second Voyager had been reprogrammed for an extended Io Volcano Watch. The spacecraft would not come closer to Io than about a million kilometers, but the geometry of the encounter was favorable for a ten-hour viewing program shortly after closest approach to Jupiter. Already, from a distance, the imaging team scientists had noted that at least three of the eight active volcanoes observed in March were still active. However, the largest of the volcanic plumes (provisionally named Pele—after the Hawaiian god, not the soccer player) seemed to have shut off entirely.

Saturday, July 7, was a leisurely day for most people at the Lab. The critical close-encounter uplink consumed a seven-and-a-half-hour block of time, which left the scientists with little to do except rest up for the Sunday night–Monday morning pass at Jupiter. Brad Smith kept busy by conducting a tour of the Lab for visiting celebrity Tony Orlando, who waved cheerfully to the unimpressed minions in Von Karman. There always seemed to be a Celebrity of the Mission —one of the inevitable consequences of being so close to Hollywood. Carl Sagan had taken Johnny Carson on a tour the day Viking 2 landed on Mars. During Voyager encounters, there were appearances by Orlando, Gregory Peck, and Angie Dickinson.

By early Sunday morning, the spacecraft was returning detailed images of the previously unseen opposite face of icy, battered Callisto. There were more "shoulder-to-shoulder" impact craters: "For people who like to count craters," said Brad Smith, "there's a lot of work there for them." Callisto seemed to be at the saturation point for cratering; there were so many that each new impact would destroy as many craters as it would create.

Sunday's pre-encounter press briefing was very low-key, as was the general atmosphere at the Lab. The impending death of Skylab and the summer-rerun mentality seemed to have muted everyone's response to the event. One journalist suggested that in honor of summer reruns the Galilean satellites should be renamed Lucy, Ricky, Fred, and Ethel. The slightly goofy mind-set that led to such suggestions was even apparent in Brad Smith's response to a question from an Italian journalist: Q—"What would you say to Galileo Galilei if he walked into the room?" A—"How were you able to live so long?"

At 1:06 AM, Monday, July 9, Voyager 2 passed within 62,000 km of Ganymede. More than 200 images were taken of the satellite's strange, jumbled surface, but most of them were recorded for playback later in the day. "We had lots of fun last night watching the Ganymede photos come in," said Hal Masursky the next morning. "They all looked like Ganymede—we're getting so we can tell a satellite from a single photo." But for most of the morning, everyone's attention was on Europa, which Voyager 1 had seen only from a great distance. Closest approach came at 11:43 AM at a distance of 206,000 km.

Europa, said Larry Soderblom, "could be the most exciting satellite in the whole Jovian system . . . it's sort of a transition body between the solid silicate [rocky] body, Io, and the ice balls, Gany-

On July 8, Europa, from 700,000 miles, showed more detail than was visible in Voyager 1 images. Dark streaks were beginning to look strangely familiar.

mede and Callisto." As the Europa pictures came in, Soderblom mused on the unexpected nature of the Jovian moons. "They're all unique, aren't they? . . . It's interesting to note how little about the system we knew. . . . We really missed—just because you know what something's made of doesn't mean you know what it *is*."

What Europa was, it became apparent, was another utterly unique world. As the pictures were displayed on the monitors in the press room, someone asked, "Wanna see my imitation of Europa?" He proceeded to smash a hard-cooked egg against a desktop—and the result was, indeed, a rather faithful representation of the Jovian moon. Other observers made the same comparison, but another analogy seemed to be in the back of everyone's mind. "It looks like some pictures of Mars I've seen," said Torrence Johnson, "but only on the walls of Lowell Observatory." Declared Hal Masursky: "The canals are back."

Percival Lowell's famous and mysterious Martian canals had finally been found—on Europa. The dark streaks seen at a distance in March now resolved into a planet-wide network of spidery, interconnected lines 10 to 50 km across. The resemblance to some of

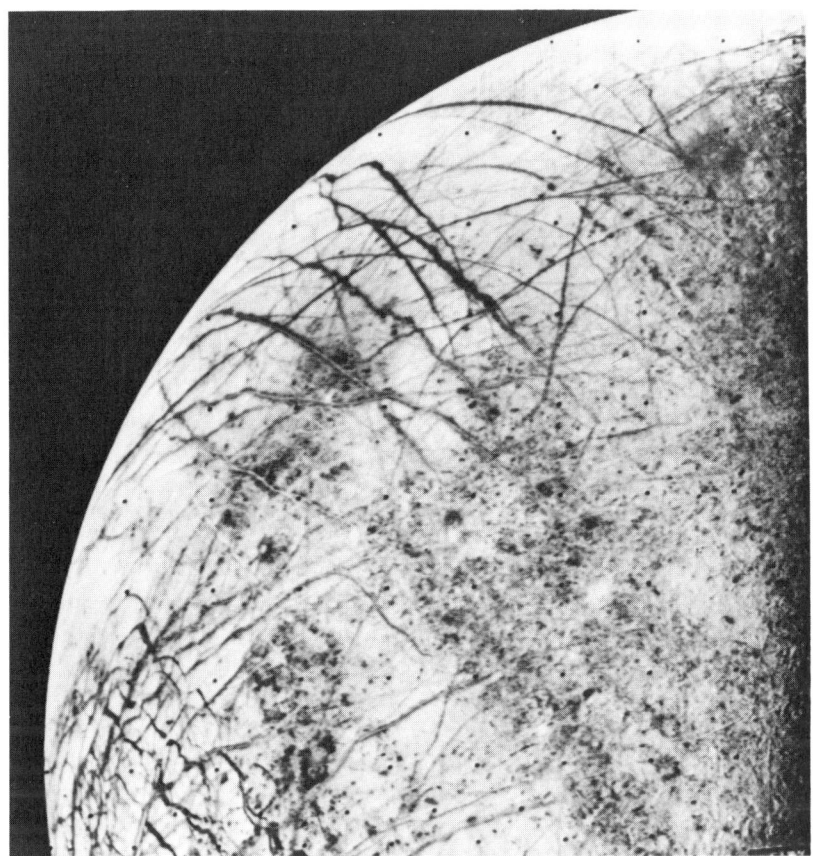

On July 9, from 152,000 miles, Europa looked astonishingly similar to Lowell's maps of Mars—complete with canals. The dark streaks were probably cracks in the surface ice. Note the absence of both relief and impact craters.

Lowell's sketches of the nonexistent Martian canals was remarkable. Even more surprising was the complete lack of vertical relief on Europa's surface. "We do not see any deep topography associated with those dark bands," said Soderblom. "If these were deep trenches . . . you'd see a bright scarp." The absence of relief seemed to eliminate the theory that Europa's surface was composed of a thin, icy shell with firm rock beneath. "We can put the thin-ice model away for another planet," said Soderblom. "It must be a warm, thick ice surface."

Beneath the invisible surface, the ice might well be in the form of a soft slush. That would account for the apparently complete lack of impact craters. "We're seeing a number of circular features," said Hal Masursky as the images were coming in. "They're small enough that it's tough to put a tag on them. . . . If you had to guess, you'd say that was a degraded crater—maybe. Io has made me a little more cautious." On Io, the volcanoes had obliterated the ancient impact craters; on Europa, it seemed, the impacts had simply been swallowed up by the slush-sea, leaving only faint, ghostlike traces of a few faded craters. Finding one craterless world had been a profound shock for the planetary scientists, but finding another, completely different craterless moon next door to it was astonishing. "We have some really nice problems to wrestle with," said Masursky.

At the noon press briefing, Ray Heacock reported that the "hardened" radiation environment was causing problems for the spacecraft. The receiver was shifting rapidly from one radio frequency to another "and we have not been able to keep an uplink continuously with the spacecraft." Commands were being sent up on multiple frequencies in the hope of finding one that the spacecraft could accept. According to Ed Stone, the high-energy electrons hitting Voyager 2 were three times more intense than anything Voyager 1 had encountered.

Larry Soderblom then mounted the Von Karman stage and showed the latest pictures of Europa and Ganymede. "Some few months ago," he said, "before the Voyager 1 encounter, we thought we had some idea of what planets were like . . . and we've discovered many times over the last couple of months how narrow our vision really was." Soderblom noted that in the gallery of planets, the Jovian satellites had shown us "the oldest [Callisto], the youngest [Io], the darkest [Amalthea], the brightest [Europa]*, the reddest [Amalthea and Io], the whitest [Europa], the most active [Io], and the least active [Callisto]. Today, we found the flattest [Europa]." The lack of relief on Europa suggested to Soderblom that instead of a thin ice crust, there might be a slushy ice mantle some 100 km thick. Soderblom also invoked the eggshell analogy: "It's as if Europa had been cracked, broken, by some process which crushed it like an eggshell and just left the pieces sitting there. Expansion and contraction of ice and water are a good way to crunch up the surface."

The tape-delayed picture of Ganymede were equally surprising. The face unseen by Voyager 1 now revealed an immense dark, circular region, later christened Regio Galileo. There were remnants of a huge ringed structure, reminiscent of the rings seen on Callisto. "This had to be enormous," said Soderblom. "The radius of curvature is very large." The rings seemed to go halfway around the planet and may have been produced by a gargantuan impact early in the

---

* As it turned out, Saturn's moon Enceladus is actually slightly brighter, or more reflective, than Europa.

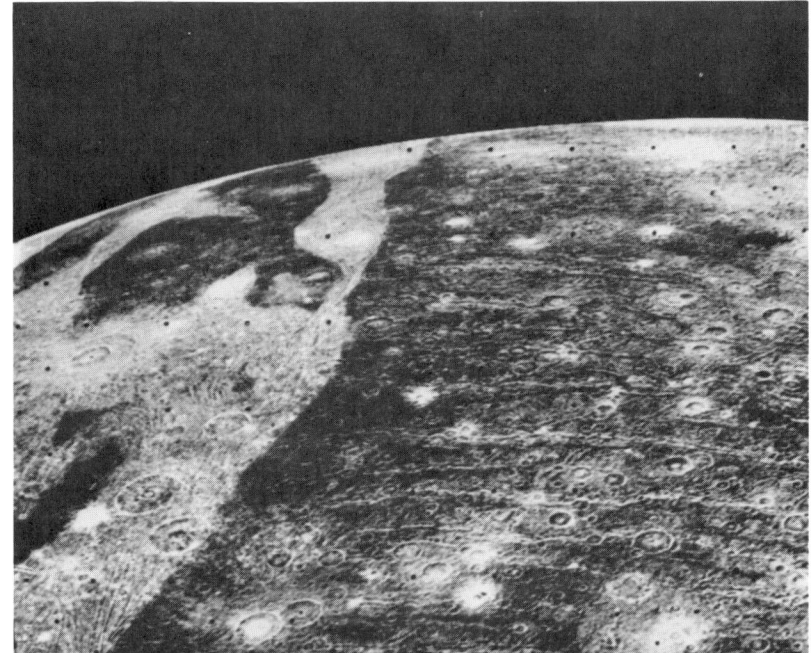

Ganymede's complex history can be deduced from these close-ups. The dark, heavily cratered terrain is probably much older than the lighter, icy patches. Faint "ripple" marks are similar to those seen on Callisto and may be remnants of a very large impact early in the planet's history.

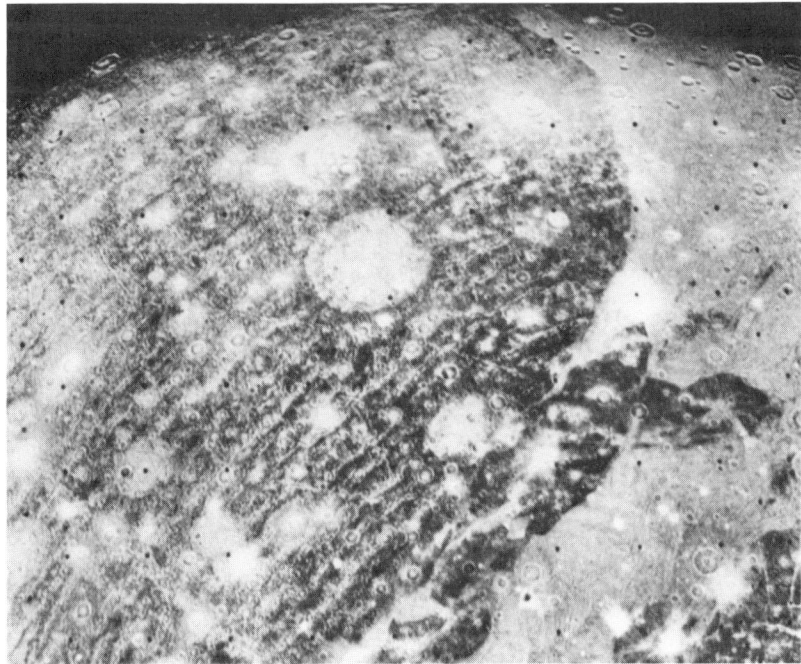

planet's history. Such an impact might have been the original cause of the weird faulting and shifting that dominated the surface of Ganymede. Such an impact, suggested Torrence Johnson, "could have turned Callisto into Ganymede."

That afternoon, at 3:29 PM, Voyager 2 made its closest approach to Jupiter, 650,000 km, at a speed of about 45,000 mph. Half an hour later, the traditional post-encounter press briefing was held. The words to remember came this time from the new NASA associate administrator for space science, Thomas A. (Tim) Mutch.

"You have been witnessing a truly revolutionary journey of exploration," Mutch told the assembled reporters. "And I will submit that you know it, but you're a little embarrassed to say it. . . . We have gone beyond the familiar part of the solar system to objects that are so exotic that their very existence, at least as far as I'm concerned, was something I'd accepted intellectually, but didn't really accept in an immediate sense. We're starting out . . . on a new stage of space exploration—on our own long journeys beyond the solar system to distant lands. . . . It's statistically unlikely that we're at a turning point in history; but if you look back at history books, such events are clearly read into the record. And I submit to you that when the history books are written a hundred years from now, two hundred years from now, the historians are going to cite this particular period of exploration as a turning point in our cultural, our scientific, our intellectual development."

Mutch was right; many of the journalists were embarrassed to admit that they had just witnessed an event of historical importance. But Tim Mutch was never a man to be embarrassed by the emotional appeal of exploration. As leader of the Viking lander imaging team, he had displayed an infectious, gee-whiz enthusiasm during the exploration of Mars; now, as a NASA official, he was unafraid to speak from the heart. Perhaps better than anyone, Tim Mutch understood the joys and imperatives of exploration—and the price. In October, 1980, shortly before Voyager 1 reached Saturn, Tim Mutch was leading a mountain-climbing expedition in the Himalayas when he was killed in a fall. It was a horrible shock to his many friends in the space science community and the press corps; but somehow, the manner in which Tim Mutch died underscored the values for which he had lived. In an unprecedented tribute, NASA administrator Robert Frosch officially renamed the Viking Lander 1 site the Thomas A. Mutch Memorial Station. A plaque was prepared, and Frosch charged his remote successors at NASA with the responsibility of delivering the plaque to Mars for installation on the base of the Viking Lander. Ironically, the plaque and the tribute probably *would* have embarrassed Mutch.

INSTANT SCIENCE

Tim Mutch had put the Voyager 2 Jupiter encounter in historical perspective, but at least one man at Von Karman that day didn't need to be told. Jurrie van der Woude, who ran the PIO picture desk, found an elderly man from Pasadena among the hundreds of reporters. The man had no press pass and had somehow managed to simply walk in off the street. He wanted to be at JPL, he said, because "this is the best show in town. Skylab, energy, gas . . . this is the only positive thing happening."

While the project officials were congratulating themselves in Von Karman, the spacecraft was going through one of the most critical maneuvers of its entire flight. As if on cue, the radio receiver began misbehaving just as the seventy-six-minute Saturn trajectory burn started. But the spacecraft was programmed to operate without instructions from Earth, and the burn went off as planned, as did the ten-hour Io Volcano Watch that followed it. The next morning, Deputy Project Manager Ek Davis reported that contact with the wayward radio receiver had been re-established: "We had to chase it around most of the night. It was a rather sweaty time for some of the managers. There were a lot of hands on, a lot of turning of the knobs."

At the briefing on the 10th, Andy Ingersoll discussed the Jovian atmosphere. "The challenge," he said, "is to understand the bandedness of the Jovian atmosphere, and the longevity. . . . At first, Voyager seemed to emphasize the chaos, not the order." Yet Jupiter *is* orderly, Ingersoll insisted. The question was "how this large-scale order could exist in the face of all the small-scale chaos . . . but I think we're beginning to recognize the order underneath."

The breakthrough, according to Ingersoll, "came as we laboriously measured the velocities of spots (small ones) as a function of latitude, and compared that with the record of Earth-based astronomy. . . . If you take all the Earth-based measurements over seventy-five years, you find that every current . . . is visible in one ten-hour sequence of Voyager pictures. They're all there!" Reta Beebe (one of a very small handful of women scientists in the Voyager project) found a "regular alternation of eastward and westward jets" that seemed symmetrical to the equator. This pattern persisted in spite of color changes and the varying appearance of the clouds. "This pattern may persist to considerable depths," said Ingersoll.

The discovery of regularity and symmetry in the Jovian atmosphere at last gave the scientists a potentially useful "handle" on the unseen interior. Brad Smith's football game analogy is useful here. A series of aerial photographs of a football game would first seem chaotic to anyone unfamiliar with the sport. But a detailed examination of the photos would turn up signs of regularity and order. The offensive formations at the line of scrimmage would vary from play to play, for example, yet it would be observed that only certain players with specific numbers on their backs would normally touch the ball. The data would show that there were two opposing teams, eleven on a side, and that they directed their energy toward moving in a particular direction. Given enough photos and analysis, an observer might come to a relatively detailed understanding of the game of football. The difficulty would come in sorting out the relevant details (the players and the ball) from the irrelevant distractions (cheerleaders, coaches, cameramen, etc.).

Ingersoll cautioned, however, that the discovery of regularity "does not provide an explanation of why there should be this pattern." The football photos, in other words, could tell observers nothing about the invention of the forward pass or why the field was 100 yards long. For Jupiter, the answers would probably have to come from a study of the deep Jovian interior. "The story will continue," Ingersoll promised.

Brad Smith reported on the Volcano Watch, which had been highly successful. Some spectacular pictures of volcanic plumes had been obtained, and it seemed that at least six of the original eight volcanoes seen by Voyager 1 were still operating. Only Pele, the largest of the eruptions, was known to be quiet. That volcano, at the center of the "hoofprint," seemed to have shut down, but not before it had wiped out the hoofprint. The notch in the flow pattern had disappeared, resulting in an oval shape surrounding the central peak. Aside from the effects of seasonal dust storms on Mars, no major topographical changes had ever been observed before on another planet.

Smith and his imaging team were still trying to make sense of Europa. In addition to the network of dark streaks, closer inspection revealed a number of very thin, bright white lines running along the center of some of the streaks. The presumption was that these white

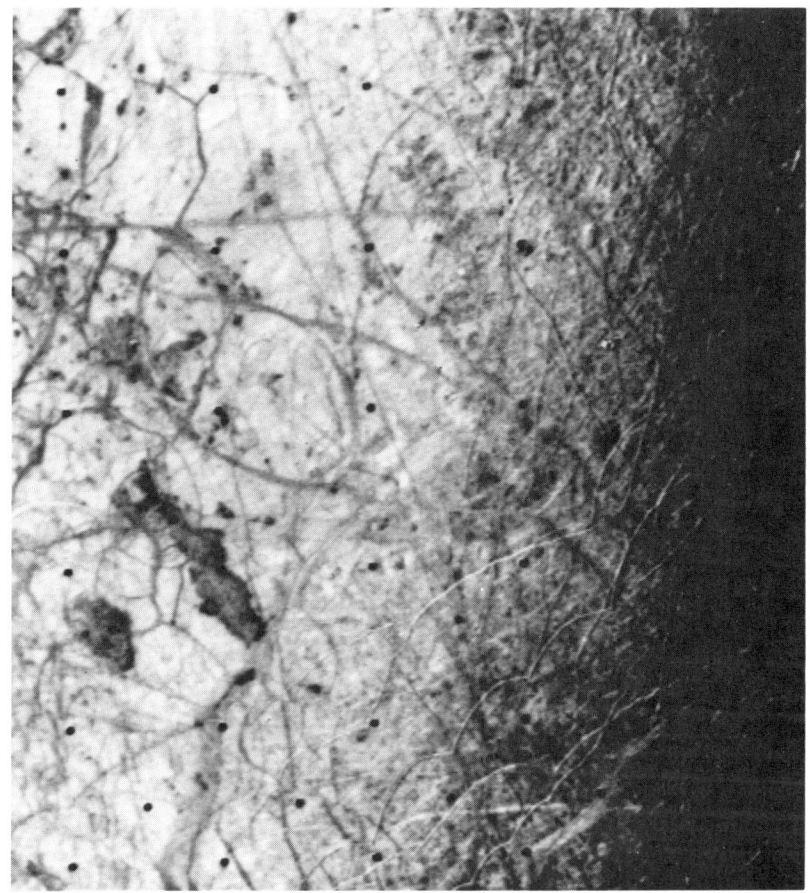

Bright streaks of fresh ice, 5 to 10 km wide, crisscross the surface of Europa. Note "scalloped" ridges running from lower center to right edge of photo. These strange features have defied explanation.

lines consisted of fresh ice that was somehow being pushed to the surface. "We call the dark areas cracks," said Smith, "but the only topography we see is in the white lines. The dark lines might just as well have been drawn on with a felt-tip pen." There seemed to be some small vertical relief associated with the white lines, perhaps no more than 50 to 100 meters. But the real mystery was the discovery of a series of long, scalloped ridges, 5 km wide and hundreds of kilometers long. These thin white lines were almost geometrically precise, and there was no obvious explanation for their existence.

That evening, Voyager 2's cameras swung around to look at Jupiter from the night side. The huge gas giant appeared as a thin, bright crescent—and just as bright, astonishingly, was the thin Jovian ring. The particles comprising the ring turned out to be excellent "forward scatterers" of light. That is, they tended to deflect sunlight ahead, rather than reflecting it backward. The forward scattering implied a small size for the particles.

The ring images showed that there was some structure in the Jovian ring, although nothing to compare with the structure in Sa-

Scientists were surprised by the brightness of Jupiter's ring when seen from the night side. The brightness implies extremely small particles in the rings, which tend to scatter light forward.

turn's rings. The main band of the ring was about 5,800 km across, but it was surrounded by dimmer material which seemed to extend all the way down to the visible cloud tops of the planet. Here was yet another new phenomenon that would have to be explained.

On Wednesday, July 11, the final press briefing was held. This was the traditional end-of-encounter science briefing—Instant Science. The principal investigators held forth on the knowledge gained by their instruments and the possible meaning of it all, but it was really too soon to integrate all the new data with the Voyager 1 and ground-based observations. Larry Soderblom, speaking about the Galilean satellites, summed up the overall situation: "We're in a relatively high state of ignorance."

Mass press conferences for the presentation of newly discovered scientific data are unique to the space program. In other fields, scientists rarely "go public" until their results have been checked and rechecked. A scientific paper submitted to a journal for publication may be delayed for months while the paper is refereed by a group of other scientists, some of whom may hold diametrically opposite viewpoints. The structure of modern science is designed to be self-correcting; and although the system may at times be sluggish and overcautious, nevertheless, it works.

Months or years after a planetary encounter, long after the space gypsies have abandoned Von Karman Auditorium, planetary scientists publish their findings in journals such as *Icarus, The Journal of Geophysical Research, Science,* and *Nature*. Frequently, these detailed studies confirm the first seat-of-the-pants impressions from the encounters; sometimes, the reverse is true. Once a year, the planetary scientists get together in person to thrash out their differences and exchange new ideas. These annual meetings of the Division for Planetary Sciences (DPS) of the American Astronomical Society are like space-science smorgasbords, featuring something for everyone.

The DPS meetings in 1979 (St. Louis), 1980 (Tucson), and 1981 (Pittsburgh) were especially bountiful, like state fairs following bumper crops. The mass of new data was enormous, and it was literally impossible for anyone attending those conferences to keep informed in every area. While papers about Mars or Venus were being presented in one room, equally interesting papers about Titan or the Jovian atmosphere were being read in another room. Debates and discussions were carried on in hotel hallways, around the Ramada pool in Tucson, and (of course) in the hotel bars.

These conferences are especially important to younger scientists. They have the opportunity to meet their professional colleagues and establish themselves as members of the space science community. For a young scientist with a freshly minted Ph.D., presenting a paper for the first time is almost a rite of passage, and it can be a highly emotional experience. At the St. Louis meeting, one young Mars expert paced the halls outside the conference room, nervously rehearsing his five-minute presentation. A friend tried to calm him down by pointing out that he would be just one out of dozens of scientists who would be presenting papers. "Yeah," he said, "but I'll be the only one to throw up."

The Voyager encounters gave the planetary scientists a great deal to talk about, but few definitive answers about anything. The Jovian system—planet, magnetosphere, ring, and satellites—turned out to be far more complex than anyone ever expected. An understanding of any one aspect of the system hinges on an understanding of everything else. The magnetosphere is incomprehensible without reference to the volcanoes of Io; the volcanoes operate by some mechanism which must be powered by the gravitational influence of Jupiter; the Jovian clouds may be colored by sulfur from Io's volcanoes, or by a complex internal process; the internal processes must be the result of the internal structure of the planet, which, in turn, depends on the early history of Jupiter; the history of the Jovian system must be deciphered from the evidence on the surfaces of the Galilean satellites; and the satellites are influenced by the magnetosphere. It all ties together—somehow. By analogy, imagine attempting to understand the Earth—with its skidding continents, evolving atmosphere, and complex biochemical interactions—on the basis of the data accumulated by a mere dozen experiments over the course of two brief flybys. Three years after the last Jupiter encounter, scientists were still trying to make sense of the Jovian system; twenty years later, they will probably still be trying.

The Jovian magnetosphere remains a challenge for physicists, but they believe they are on the right track. "I don't think we have a fundamental problem," said experimenter Fred Scarf. "We have lots of details that have to be studied."

The general size and shape of the magnetosphere are now known. The magnetosphere bulges out along the magnetic field lines close to the planet, then gradually spreads out into a thin disk aligned with the Jovian equator. Out to a distance of about 20 R_J, the charged particles in the magnetosphere rotate with the Jovian magnetic field. Beyond that distance, the particles lag behind the spinning planet; but at a still greater distance, the co-rotation re-establishes itself. Some of the charged particles are accelerated to extremely high velocities and spurt out of the magnetosphere into the interplanetary medium. The Voyagers detected these particle streams as far as 800 R_J from the planet; there is some evidence that these Jovian emissions may extend their influence all the way to Earth. It is possible that some Earth auroras are produced by Jovian rather than solar particles. In the opposite direction, away from the sun, the magnetosphere turns into a long, extended magnetotail, a kind of Jovian shadow. The magnetotail is millions of kilometers long and may reach the orbit of Saturn. In fact, scientists were optimistic that Voyager 2 might catch Saturn within the Jovian magnetotail.

Voyager 1 discovered an incredible hot spot inside the magnetosphere near its outer boundaries. The plasma in that region was estimated to be some 300 to 400 *million* degrees centigrade—compared with a presumed 20-million-degree temperature in the core of the sun. Voyager 1 passed right through this area with no ill effects because the hot region is also the most nearly perfect vacuum yet observed in the solar system. With only one particle for every hundred cubic centimeters, the hot plasma posed no threat to the spacecraft.

Although this plasma is very tenuous, it is also extremely energetic. It may be the controlling factor in the pulsations of the magnetosphere observed by both Voyagers. Basically, the pressure of the plasma forms an outer shield against the incoming solar wind. If the solar wind speed varies, the balance between the opposing forces is disrupted, plasma may be ejected from the field, and the outer magnetosphere collapses. Newly accelerated particles injected from the inner magnetosphere re-inflate the field in a matter of hours.

The major source of charged particles in the magnetosphere seems to be the surface of Io. The high-speed sulfur and oxygen ions observed by the spacecraft are produced either by material "sputtered" into space by the impacts of tiny particles or by the sulfur dioxide or hydrogen sulfide plumes from the volcanoes. The apparently ceaseless eruptions, the sputtering, and possibly some other processes provide a constant source of replenishment for the Io torus and the surrounding magnetosphere. The plasma must be renewed because some of it escapes the magnetosphere in high-energy bursts, and some of it is absorbed by the Galilean satellites as their orbits take them through the plasma clouds. Some estimates suggest that Io must be supplying the magnetosphere with about two tons of matter every second.

Some of the particles from Io escape the plasma torus by spiraling inward along the magnetic field lines until they hit the upper atmosphere of Jupiter in the vicinity of the poles. There, the collisions produce auroras, which were observed by both Voyagers. The influx of sulfur and oxygen ions may also have an influence on the color and composition of the Jovian clouds. Sulfur, in its various forms, may act as a coloring agent, while the oxygen may combine with carbon to form the carbon monoxide detected in the atmosphere. The energy involved in these Io–torus–Jupiter interactions is enormous, on the order of one to two trillion watts.

The Jovian atmosphere may be influenced by external contributions from Io or even tiny Amalthea, but the key to an understanding of Jupiter's swirling clouds seems to be hidden in its interior. The

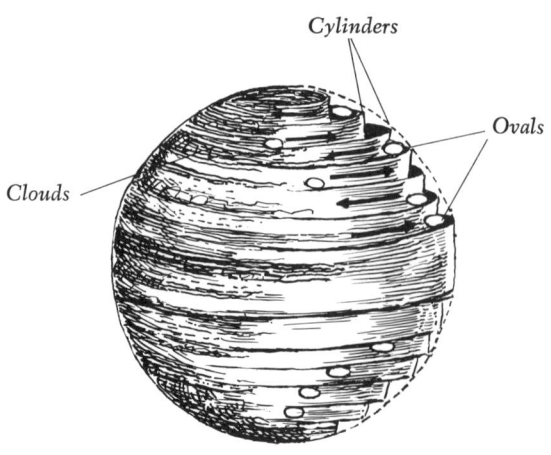

The Cylinder Model of Jupiter's Atmosphere

Voyagers directly observed only the cloud tops and (through upper-level holes) some intermediate cloud decks. The lateral motions of the clouds have been plotted in detail, but the unseen vertical motions within the atmosphere still baffle planetary scientists. The lateral motions must be tied to the vertical motions, but at our present level of knowledge it is still possible to explain the visible motions of the clouds with utterly different conceptions of Jupiter's interior.

To Andy Ingersoll, the zonal flows look like atmospheric manifestations of the deep internal structure of the planet. The persistence of the flows and their symmetrical pattern relative to the equator seem to imply a Jovian interior that consists of a series of rotating "nested" cylinders. The deep interior of the planet, including the metallic hydrogen layer and the dense core, would be incorporated into the spinning cylinders. Where the cylinders intersect with the surface of the sphere, the visible zonal flows simply reflect the interior motion. A flow at, say, 50° north latitude is virtually identical to a flow at 50° south because they are opposite ends of the same cylinder. "Such behavior is not as unlikely as it first seems," Ingersoll argues. Trying to explain his theory to a group of dubious reporters, he confessed, "I hate to say, 'It's in the equations,' but it is."

It's not in everyone's equations, however. Planetary scientist Gareth Williams of Princeton has produced a computer model of the Jovian atmosphere without resorting to nested cylinders or, indeed, to any deep interior process. In Williams' equations, the zonal flows and eddies can be explained completely in terms of the behavior of a thin upper atmosphere layer no more than about 100 km deep. The eddies feed their energy to the zonal flows, and the deep vertical transport of energy is unnecessary. Since both models produce ac-

A model proposed by Andrew Ingersoll of Caltech suggests that the interior of gas giants may consist of "nested" oppositely directed cylinders of rotating gas. The atmospheric bands (belts and zones) would thus be a surface manifestation of a deep interior process. Competing models suggest that banding is a near-surface phenomenon which does not involve the deep interior.

curate representations of the observed atmospheric motions, it is difficult to choose between the two.

The exact composition of the Jovian atmosphere also remains uncertain. The present model is based on the "solar composition" argument and assumes the presence of some compounds that have not been directly observed. Hydrogen sulfide and water have both eluded our detectors, possibly because they precipitate out at too low a level in the clouds. On the basis of both theory and observations, it seems that there are three distinct cloud layers in the upper atmosphere. The bottom layer ought to be composed of either water ice or liquid water droplets. Above that, there should be a layer composed of ammonium hydrosulfide (NH_4SH)—a combination of ammonia (NH_4) and hydrogen sulfide (H_2S)—and a topmost layer of ammonia ice.

Holes in the upper clouds have let us see to a level where the temperature is at least 300° Kelvin (or 27°C; 0°K—absolute zero—equals −273°C) and both water and hydrogen sulfide ought to be abundant. Ingersoll suggests that these hot spots may be the Jovian equivalent of deserts, where "dry" air that has already lost its condensable gases is descending back to the interior.

The colors of the Jovian clouds are not only beautiful but revealing. Voyager and Earth-based infrared observations correlate well with the observed cloud colors, so cloud color seems to be a good indicator of altitude. We see deepest into the atmosphere where the clouds are blue. Above the deep, hot blue regions come the brown clouds, followed by white and then red. "The trouble is," writes Ingersoll, "that all cloud species predicted for equilibrium conditions are white. Color must come when chemical equilibrium is disturbed, either by charged particles, energetic photons, lightning, or rapid vertical motion through different temperature regimes." If the Jovian atmosphere were perfectly stable, it presumably would be colorless. The precise coloring agent is still unknown; it might be sulfur from Io, phosphorus from the Jovian interior, or organic molecules manufactured by photochemical reactions within the atmosphere.

At some point, virtually every theory of interior structure, atmospheric circulation, cloud color, and temperature runs afoul of the Great Red Spot. The Spot has been observed for more than three

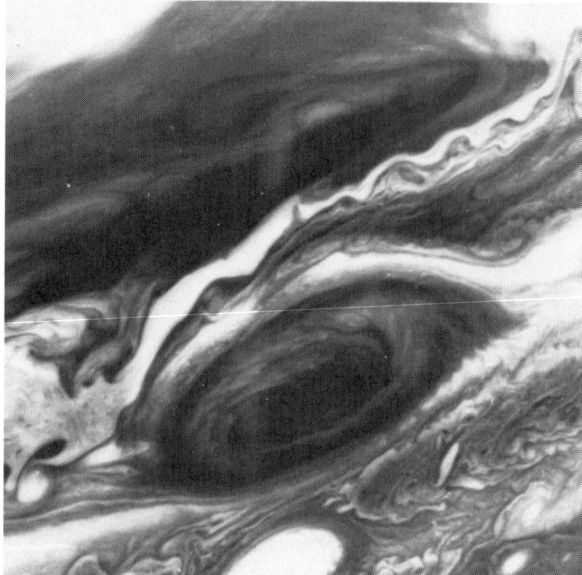

Flow patterns around the Great Red Spot are a little less chaotic in Voyager 2 images, but the complexity of the atmospheric flow is still apparent. Note the structure within the Spot itself.

hundred years, so it cannot be dismissed as a temporary anomaly in the normal flow patterns. It is clearly a member of a family of objects, including the large white ovals (which were observed forming around 1939) and the lesser red and white spots, but its great size and longevity make it unique. Several theories offer explanations for the Spot, but each rests on assumptions about the interior which cannot be proved. For planetary scientists, the Great Red Spot remains as compelling and infuriating as Ahab's Great White Whale.

The spots, great and small, may be similar to hurricanes, drawing their energy from convection cells transporting heat from the interior. They may be upper-atmosphere phenomena, extracting energy from the zonal jets. Or they may *feed* the zonal jets. Or they may be cannibals, maintaining their energy by gobbling up smaller spots. Planetary scientists trying to understand the Spot are taking a closer look at flows and eddies in terrestrial oceans, weather patterns in our own atmosphere, and the subtle details of wave interactions. Nevertheless, after two Pioneer and two Voyager encounters, the Great Red Spot is still unexplained. Someone may yet come up with a brilliant synthesis of the existing data; more realistically, it seems likely that the Spot will remain mysterious for years to come.

What is needed are data from *within* the Jovian clouds. A simple atmospheric entry probe could unlock the secrets that were hidden from Voyager's view. Such a probe is being built, but whether it will ever actually fly is anyone's guess.

NASA's choice for a post-Voyager Jupiter mission is known as Galileo. The mission would combine a Voyagerlike Jupiter orbiter with a short-lived atmospheric probe that would descend beyond the level where the water-ice clouds are presumed to be. Below that level, the atmospheric pressure will crush the probe and the intense heat will fry it. During its lifetime of about an hour, the probe could collect an enormous amount of valuable data on atmospheric composition, wind speeds and direction, and pressure and temperature. It could answer numerous questions about heat flow and would give us new parameters for theories of the interior. The Jupiter orbiter, meanwhile, would survive for about two years and give us multiple close passes at several Jovian moons.

But the Galileo mission hangs by a slender thread. It was delayed for several years by problems with the launch vehicle (the Space Shuttle) and by penny-pinching in Congress. At present, Galileo is scheduled for launch in the mid-eighties, and arrival at Jupiter in early 1990. It would be an exceptionally valuable mission, but it may never get off the ground. Galileo has survived several rounds of budget cutting in the Carter and Reagan Administrations, but it may become a prime target for the Reagan budget cutters. If Galileo is cancelled, it is doubtful that we will be returning to Jupiter in this century. The questions raised by Voyager will be left for our grandchildren to answer.

Our post-Voyager view of the Galilean satellites is also incomplete and confused. Before the Voyager encounters we knew almost nothing about these four bodies. Most planetary scientists expected that they would be rather dull to look at; Ganymede and Callisto would be bland ice balls, Europa would look like our moon, and Io would look like a red version of it. In fact, the Galilean satellites turned out to be four absolutely unique bodies, as different from each other as they are from everything else in the solar system. "I don't think we could have been more wrong in predicting what we would see on the Galilean satellites," Brad Smith has written. "What we had failed to consider is that ice becomes as rigid as steel in frigid space five astronomical units from the sun; also, we could not have known that there were processes occurring within Io and Europa that had not yet been encountered in planetary exploration."

Taken as a group, the four moons seem to confirm the standard picture of a hot, incandescent young Jupiter. The Jovian system apparently mimicked the young solar system, with the volatile gases and

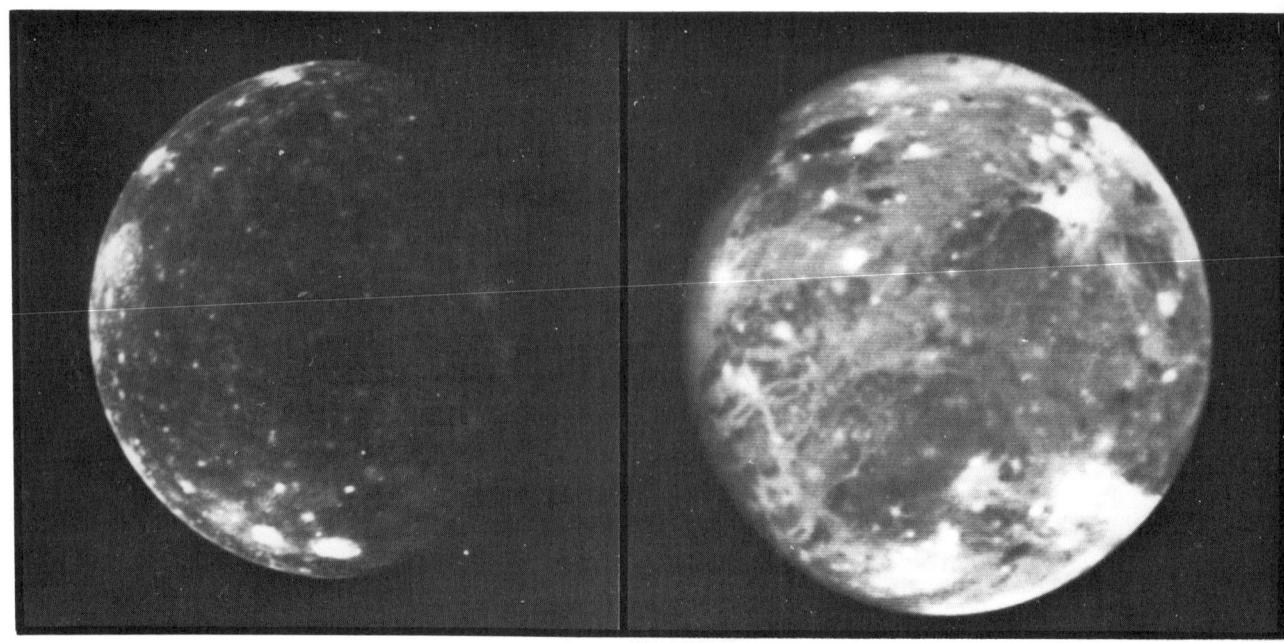

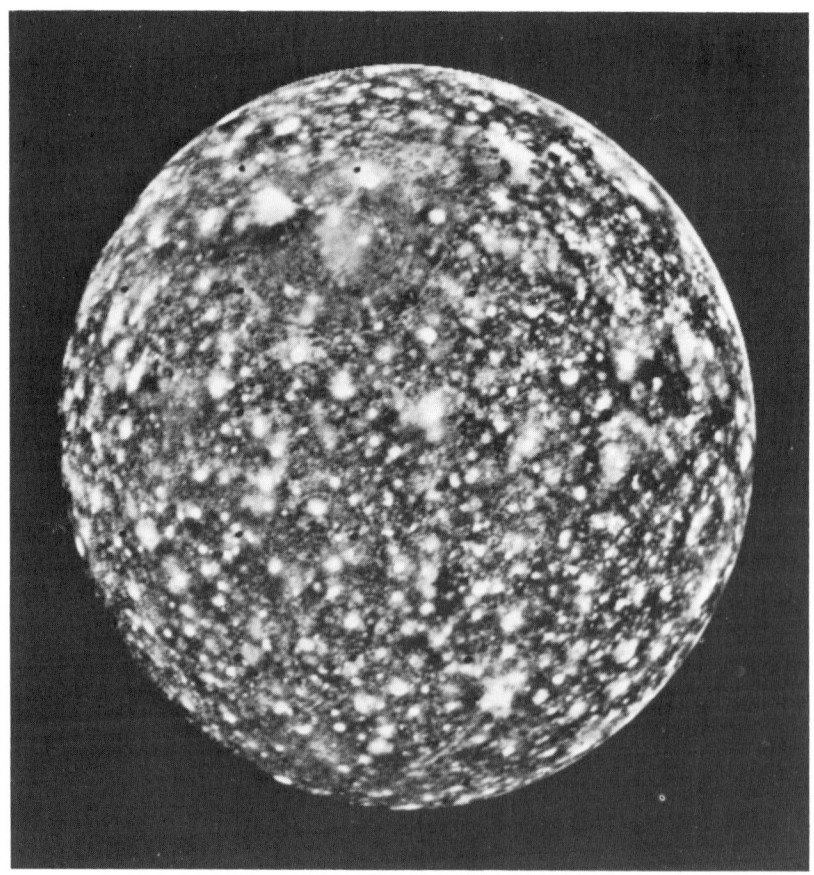

Callisto may be the most heavily cratered body in the solar system. This enhanced image shows craters at the saturation level—each new impact destroys as many craters as it creates.

A rogue's gallery of mysterious moons (left to right): Callisto, Ganymede, Europa, and Io, each shown to actual scale. Before Voyager, they were no more than dim points of light. Now we know them as individual worlds, each with its own unique history, each harboring its own challenges for Earthly scientists.

water being "cooked" out of the inner bodies while the outer bodies retained icy surfaces and a lower density.

Callisto is the most familiar-looking body, a crater-saturated ancient surface. Most theories of the origin of the solar system suggest that the ancient epoch of heavy cratering ended around four billion years ago; the surface of Callisto, a geologically dead world, has preserved the record of that early bombardment. This is just the reverse of what had been expected; most scientists believed that the cratering record would be found on dense, rocky Io, while the icy surface of Callisto would have obliterated most of the early impact craters.

If, at first glance, Callisto looks like the Earth's moon or Mercury, closer analysis turns up some fascinating differences. On our moon, Mercury, and Mars, rocky bodies which have or once had hot, molten interiors, the regions of heavy cratering are punctuated by large "plains" and basins, where huge ancient lava flows filled in the depressions created by very large impacts. On Callisto, no such basins are visible. There is also a conspicuous lack of craters larger than about 150 km in diameter. Another difference is the structure of the craters, which are flatter than their lunar counterparts and don't have central depressions or mountain rings surrounding them. In fact, there do not appear to be mountains of any sort on Callisto.

All this leads geologists to conclude that Callisto has a rather thin crust composed of an ice–rock mixture. Below the crust may lie a thick mantle of ice or, if internal temperatures are high enough, possibly water. A very large impact might fracture the crust deeply enough for the softer mantle to influence the surface. The large impacts would be deformed by the slowly moving ice below. The concentric ringed structures remaining on the surface are probably the frozen remnants of these early, giant impacts. Smaller impacts would be preserved in the stiffer upper crust.

Parts of Ganymede look a great deal like Callisto: the puzzle is why the rest of it looks so different. Ganymede and Callisto are the closest things to twin planets the solar system has to offer—similar size, density, and composition, and the same planetary address. But very early in their histories, Callisto and Ganymede started down different roads.

The dark, heavily cratered regions on Ganymede, such as Regio Galileo, seem to be the oldest and most Callisto-like. But the jumbled blocks of lighter material and the grooved terrain complicate the

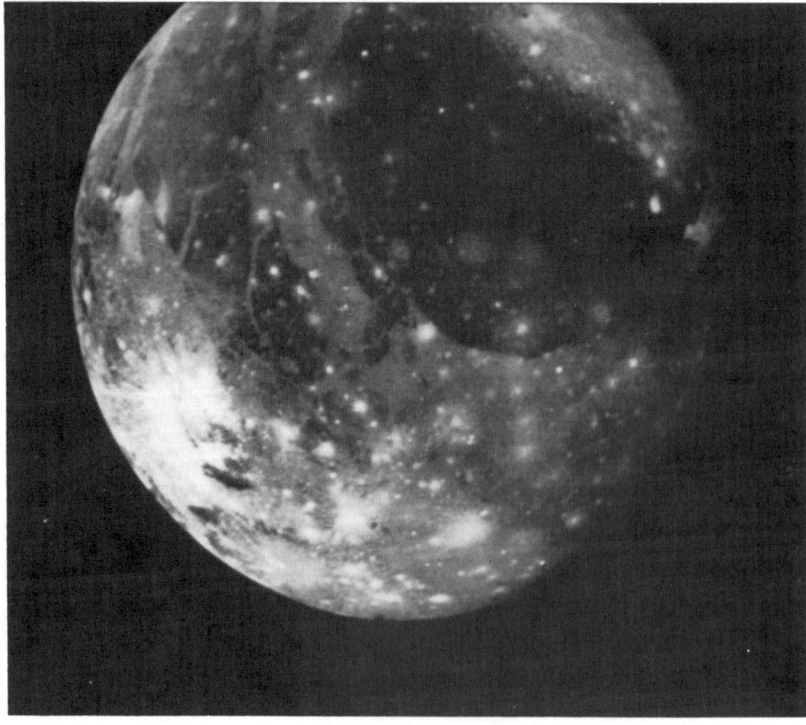

Dark and ancient, Regio Galileo is some 2,000 miles in diameter. This may be the site of an immense impact which could have started Ganymede's crust in motion, leading to the moon's present-day crazy quilt appearance.

picture enormously. Some groove systems overlie or bisect the ancient cratered terrain; at other places, craters and lighter surface material have covered the grooves.

It seems probable that on Callisto, farther from the hot proto-Jupiter, the crust cooled and stiffened earlier than on Ganymede. With its crust maintaining its plasticity longer, Ganymede had more time for geological convulsions such as faulting and shifting. It is also

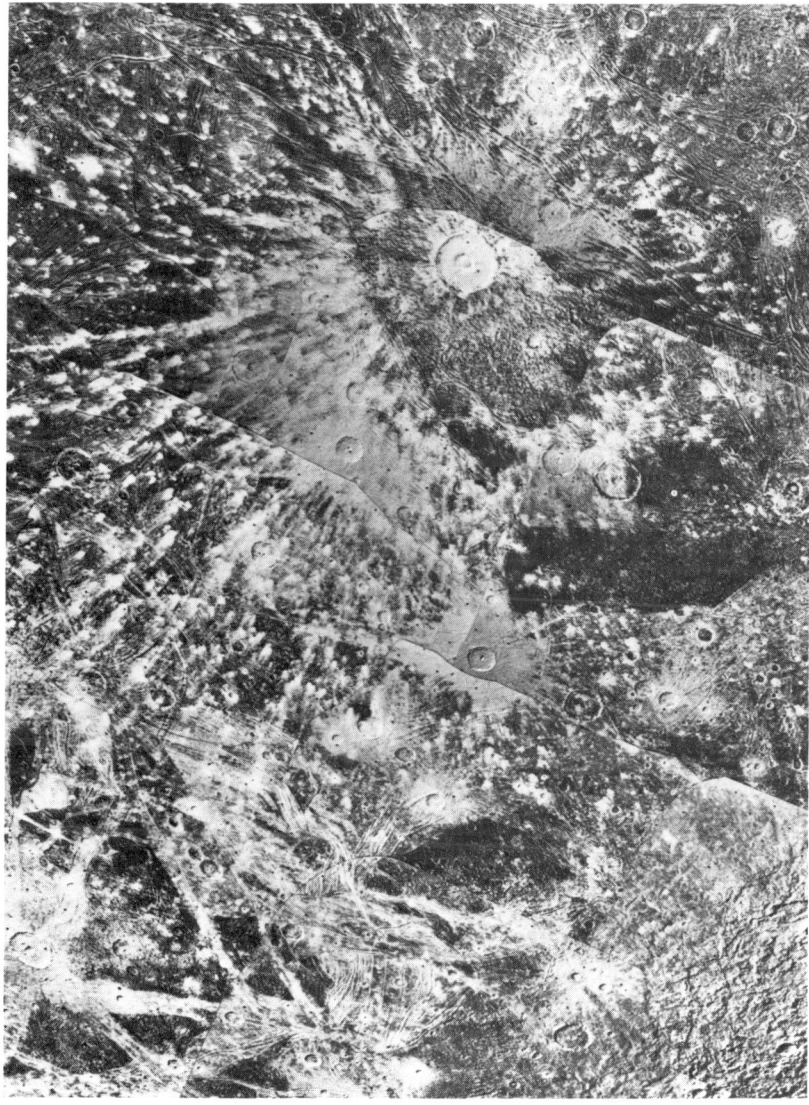

This photomosaic reveals the challenge facing geologists attempting to explain the surface of Ganymede. The large rayed crater at top is 93 miles in diameter.

possible that Ganymede's interior produced more heat than Callisto's because Ganymede is larger. There is also a chance that Ganymede's surface was set in motion by an immense impact early in its history. All these processes may have played a role in transforming the surface of Ganymede, but no one has yet come up with a tidy theory that explains everything.

It does seem clear that whatever happened on Ganymede, it happened early. Even the grooved terrain is probably at least three and a half billion years old. Possibly, the grooves formed just as the crust was becoming rigid; craters on top of the grooves are less deformed than craters in the dark, ancient terrain. Ganymede's crust was able to support some relief, unlike Callisto, so Ganymede has some modest mountains about 1,000 meters high, roughly equivalent to the Appalachians on Earth. The ice–rock crust, resting on a water or ice mantle, probably can't support more massive mountains.

Ganymede may ultimately tell us more about the Earth than any other planetary body. Large-scale tectonic activity had never been observed anywhere else but our home planet; finding it on Ganymede, a very un-Earthlike world, was a fantastic surprise. Yet Ganymede's crustal motions seem to operate on different principles from Earth's. Here, new material is brought to the surface along the midocean ridges, and the huge continental plates sink back into the interior in regions known as subduction zones. Ganymede seems to possess no analogous structures, and the visual evidence suggests that Ganymede's crustal dance was much more chaotic than Earth's.

Europa, the unexpected star of the Voyager 2 summer rerun, may be the most inexplicable of all the Galilean satellites. Ganymede and Callisto have obvious similarities to other planets and Io, bizarre though it may be, seems to behave comprehensibly. But Europa, the cracked eggshell, stands alone, a one-of-a-kind world that seems to have escaped from someone's science-fiction novel: Iceworld, planet of a thousand mysteries.

Based on Europa's density (3.0), scientists calculate that if all of Europa's water was brought to the surface during the era of the young, hot Jupiter, the planet could have an ice crust some 75 to 100 km thick. There is no way to be certain of this, however; some models suggest that the ice crust might be no more than a few hundred meters thick.

Careful analysis of the Voyager 2 high-resolution images revealed a few ringed features that seem to be old, degraded craters. Most of them are only about 20 km in diameter, and none is larger than about 50 km. The scarcity of identifiable craters would normally indicate extreme youth for the surface of Europa; but if it is assumed that the cratering rate at Europa was only one-tenth as high as for Earth's moon, then Europa's surface might be as old as three and a half billion years. But although there are plausible reasons for assuming a lower cratering rate (Jupiter might act as a shield, for example), no one can say with any certainty whether this was actually the case. If the cratering rate was normal, then Europa's surface may be younger than five hundred million years. If the younger age is accurate, then there is probably an on-going resurfacing process at work.

That resurfacing process may be visible in the thin white streaks that are present at the center of some of the dark "cracks." The white ridges may be fresh ice from below, being squeezed upward to the surface through weak spots in the crust. Calculations show that the crust itself may have expanded by as much as five percent, probably as the result of the freezing of a primordial Europan ocean.

Some scientists suspect that Europa may still have an ocean of liquid water hidden beneath the frozen crust. If internal temperatures are high enough—possibly due to the radioactive decay of heavy elements or the tidal forces of Jupiter—then there may be a global subsurface ocean on Europa. Such a view is consistent with a "packed sea ice" analogy for the surface cracks, and it carries an interesting implication. The one item considered essential for the evolution of life that has never been observed on other planets is a long-lived supply of liquid water. Mars may have had abundant liquid water, but it disappeared billions of years ago; new data suggest that Venus may once have had liquid water, but it is now too hot, and Titan is too cold. But if there has been a subsurface ocean on Europa these past four billion years, the evolution of living organisms may have been possible. In the view of some, Europa may be a better place to look for life than either Mars or Titan.

With or without life, Europa remains an enigma. The long, scalloped white ridges still defy any reasonable explanation. These graceful, precisely curved lines are "so bizarre," writes imaging team scientist David Morrison, "that one tends not to believe the reality of what is seen. Nothing remotely like [them] has ever been seen on any other planet." Morrison's conclusion is depressing but accurate: "At present the geology of Europa remains beyond our understanding."

And then there is Io. This fizzing, fulminating Yellowstone of a planet doesn't just sit passively for portraits, it *does* things. No other rocky planet—excepting the Earth—has ever been observed in the act of *doing* things. Other worlds show us evidence of ancient events, performances that closed a billion years ago, but Io is a long-running show that has at last found its audience.

To say that Io is volcanically active is like saying that Baryshnikov is a dancer or Jack Nicklaus is a golfer—it doesn't quite express the true situation. On Io, there are more than two hundred volcanic features larger than 20 km; on Earth, with three and a half times more surface area, there are only fifteen. Imagine a half dozen Mount St. Helens erupting simultaneously all over the globe, continuously, with new volcanoes popping up here and there every few years, and you will have some idea of what the Earth would be like if it were as active as Io.

Io's volcanoes are not only more numerous, they are also more energetic than those of Earth. Material ejected from Io's volcanic vents has a velocity as high as 1,000 meters per second, or about ten times greater than the vent velocity of terrestrial volcanoes. Volcanic plumes as high as 300 km have been observed in Voyager images, although this is somewhat deceptive due to Io's low gravity. Geologist Sue Kieffer has calculated that if the famous "Old Faithful" geyser of Yellowstone were located on Io, its jet of steam would rise more than 35 km.

Voyager infrared measurements detected many hot spots on the surface of Io, most of them apparently associated with very dark features which could be molten rock or sulfur "lakes". The temperature of one such feature, near the volcano Loki, was measured at 17°C, or just about room temperature. The typical "noontime" temperature on Io is about −150°C, but a few small hot spots have been measured to have temperatures as high as 250°C. Averaging the hot and cold spots together, the surface of Io is emitting about 2 watts of energy for each square meter of surface area. The average

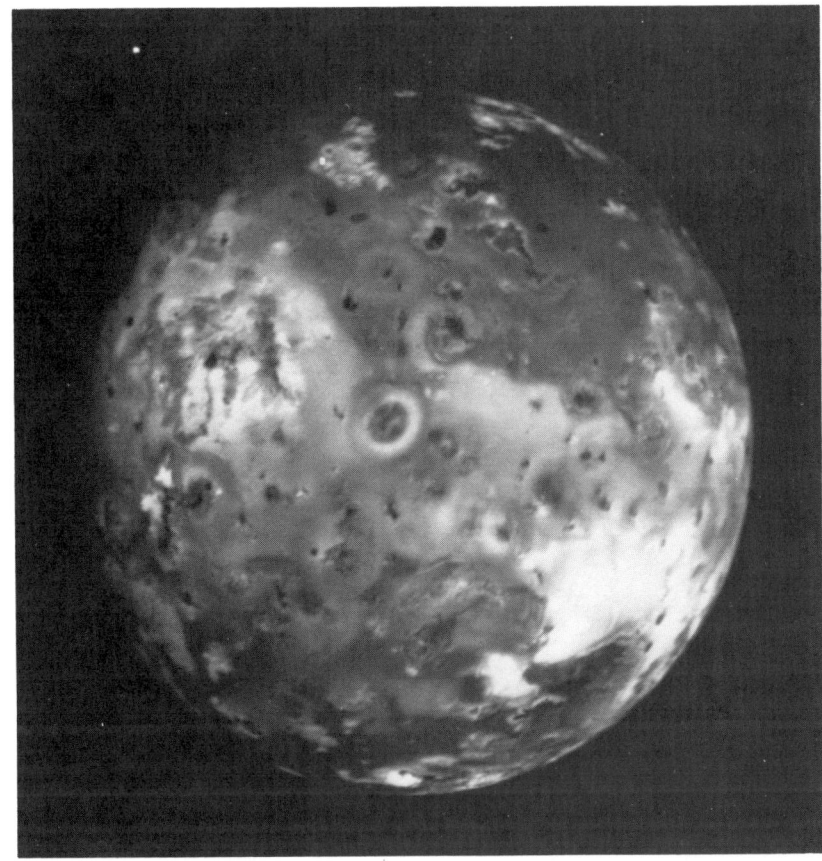

This Voyager 1 image shows the incredibly energetic surface of Io. Most of the visible features seem to be of volcanic origin. Tidal heating may be the source of Io's energy.

for Earth is about 0.06 watt per square meter. Torrence Johnson points out that even the most active of Earth's geothermal hot spots produces just 1.7 watts per square meter, or less than the *average* emission for the entire globe of Io.

Where does all that energy come from? In their famous *Science* paper, Peale, Reynolds, and Cassen proposed that Io is being heated by a gravitational tug-of-war between Jupiter and the other Galilean satellites. If Io's orbit were completely regular and uniform, there would be no tidal heating. But the other moons, particularly Europa, perturb Io's motions slightly, causing it to "nod" back and forth with respect to Jupiter. This nodding produces internal friction which heats the interior and keeps it in a molten state. However, Io is emitting so much heat that even this tidal process must be incredibly efficient. Some scientists suspect that other processes may be involved as well.

The mechanism that drives the Io volcanoes is the subject of considerable debate in the space science community. Several models have been proposed, but there are problems with each of them. No one is completely sure how volcanoes work on Earth, and trying to

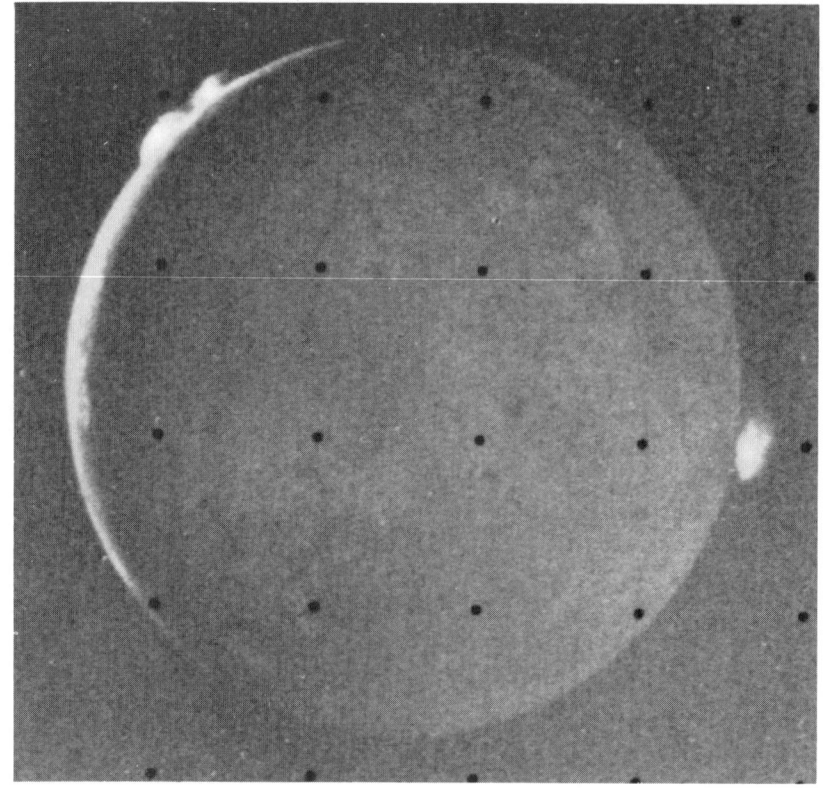

Still active, Io's intense volcanism may be resurfacing the planet at a rate of several millimeters per year. The two volcanic plumes at left are about 60 miles high; the plume at lower right is about 100 miles high and 200 miles wide.

understand so alien a world as Io by Earth-analogy may be misleading.

In one model, Io's surface is composed of a crust of solid sulfur and sulfur dioxide, resting on a second layer of solid sulfur and liquid SO_2. Below that lies a thick layer of molten sulfur, heated by silicate lavas from the deep interior. The molten sulfur ocean heats the liquid SO_2 above it, turning it to gas which expands rapidly and bursts out through the surface. Another model proposes a sulfur-enriched silicate crust. In this case, the denser upper layers slump into the sulfur ocean, resulting in heating and the sudden, violent decompression of the trapped SO_2 gas.

If the sulfur-crust model is accurate, then Io's mountains are difficult to explain. Mountains more than 10 km high have been observed; if they were composed primarily of sulfur, they would simply flow away like glaciers. On the other hand, if the mountains are composed of the heavier silicates, as on Earth, it would be difficult for the underlying sulfur to support them.

However they do it, Io's volcanoes are ejecting about 100,000 tons of material each second. The deposition of so much material on the surface keeps Io looking young and craterless. According to a new analysis by one JPL scientist, David Pieri, the resurfacing rate may be as high as eight millimeters per year. "That's ten times more than you need to obliterate impact craters," says Pieri. Over the course of geologic time, Io may have literally turned itself inside out, recycling the crust material many times over. Only a small fraction of the material is lost to the magnetosphere, although the total loss over four billion years may amount to an erosion of several hundred meters of Io's surface. As Torrence Johnson puts it, "Io seems to be in no danger of eroding away before our eyes."

However, at least some of Io has eroded away, and it is visible on the surface of Amalthea. This small potato-shaped moon seems to be coated with dark red ejecta from Io. Amalthea was long thought to be Jupiter's innermost moon; like so many other Jovian "truths," this has been changed by Voyager.

As of this writing, at least three new Jovian moons have been detected in post-encounter analysis of Voyager images, giving Jupiter a total of sixteen. Very likely, still more moons are waiting to be discovered. The newly crowned innermost moon is known as 1979J1, and its existence may help explain the Jovian ring system. The new moon orbits at the outer edge of the system and is probably the source of much of the ring material. The satellite is continually "sandblasted" by micrometeorites and material from Io and the magnetosphere. The sandblasting strips material from the surface of the tiny moon, which then becomes part of the rings. The moon's gravity is so low that the dust stripped easily escapes to space. It seems that the Jovian ring system is continuously replenished in this manner; some scientists suspect that there may also be other small moons hiding within the ring itself, each making its own contribution to the population of particles in the ring.

The new moons seemed to make Jupiter's new ring comprehensible, but the Voyager scientists were about to discover how little they really knew about planetary rings. Their two spacecraft had left Jupiter behind and were on course for the planet that was the unquestioned Lord of the Rings. The frontier of mankind's universe was about to leap a half billion miles outward, to a world that would make Jupiter seem sane and orderly, to a world of austere beauty and awesome mystery, to a realm beyond imagining . . . to Saturn.

8 PIONEERING

Pioneer 11 departed Earth on April 5, 1973. Not far from the Pioneer launch pad at Cape Canaveral, a huge Saturn V booster was being readied for the May 14 launch of Skylab. On that same day in Washington, John Dean was preparing to meet with federal Watergate investigators, and in San Clemente, John Ehrlichman was telling the judge in the Daniel Ellsberg trial that he would be President Nixon's new nominee for Director of the FBI. Spiro Agnew was the Vice President, Jimmy Carter was the Governor of Georgia, Ronald Reagan was Governor of California, and Hank Aaron was the left-fielder of the Atlanta Braves. Gasoline cost about sixty cents a gallon. By late August of 1979, Aaron had retired in glory and Nixon and Agnew in disgrace, Carter was President and Reagan wanted to be, gasoline was hard to find at any price, and Skylab was scattered all over the Australian outback. Pioneer 11 had seen a lot of history; now, it was about to make some.

After a year of Jovian revelations mankind was about to get its first close look at the planet Saturn. The most distant world known to the ancients, Saturn was the first "new" planet to be visited by a spacecraft since Pioneer 10 swung past Jupiter in December, 1973. Nearly a billion miles from the sun, eighty-six light-minutes from the antennas of the Deep Space Network, Saturn was still, as Churchill once said of Russia, an enigma wrapped in a mystery.

Pioneer 11 wasn't supposed to go to Saturn. It was designed as a Jupiter probe, twin to Pioneer 10, and its basic mission was to have lasted little more than two years. But following launch, scientists and engineers realized that by using the gravity-assist technique at Jupiter, Pioneer 11 could be targeted for a 1979 Saturn encounter. It would be extremely useful for the Voyager mission if Pioneer could provide some preliminary reconnaissance of the Saturnian system. No one had any idea of what the environment of the sixth planet was actually like; it seemed entirely possible that a Saturn encounter near the rings could be a kamikaze mission.

The Pioneer 11 flight plan was changed, and after Jupiter encounter in late 1974, the spacecraft executed a sharp U-turn and cut back across the inner solar system for a rendezvous with Saturn, which was on the opposite side of the sun. The six-year, two-billion-mile journey was the longest and most ambitious ever attempted.

Even while en route, Pioneer blazed new trails. The Saturn trajectory took the spacecraft far above the plane of the ecliptic, into totally unexplored regions. In a sense, the solar system is like Flatland, with nearly everything orbiting within the narrow confines of the plane of the sun's equator; Pioneer 11 gave us our first "three-dimensional" look at the solar system. At the peak of its long curve, Pioneer 11 was more than a hundred million miles above the plane of the planets, greater than the distance from the Earth to the sun. The out-of-ecliptic trajectory gave us previously unavailable data on the solar wind and other phenomena.

Now, the spacecraft was descending back toward the plane of the ecliptic. At Saturn, it would loop below the plane of the ecliptic, then back up and on toward interstellar space. It would cross the plane of Saturn's rings twice, once coming and once going. Crossing the ring plane would be the acid test for both Pioneer and Voyager.

If the visible rings were composed of trillions of tiny particles, how many more unseen particles were there, waiting to smash a speeding spacecraft?

There were two possible choices for the Pioneer Saturn encounter trajectory, and both looked risky. The "inside" trajectory would take the spacecraft to within just 3,600 km of the Saturnian cloud tops. The inside track would be well inside the orbit of the C ring, innermost of the confirmed rings. However, in 1969, French astronomer Pierre Guerin had found evidence of what appeared to be a dim D ring inside the C ring. Not everyone was convinced that the D ring actually existed, but if it did, Pioneer 11 would be passing uncomfortably close to it. A small gap existed between the D and C rings (the Guerin Division, also known as the French Division), and the spacecraft might be targeted to pass through it, but the risk seemed formidable.

The "outside" trajectory also looked like a gamble. The bright A ring extended out to 77,400 km from the planet, and Pioneer could be targeted to cross the ring plane at any point outside the A ring. But there was some evidence for the existence of yet another ring, the E ring, outside the A ring. If the E ring existed, the spacecraft might very well collide with it on the outside trajectory.

Presented with a choice of possible disasters, the Pioneer principal investigators bit the bullet and voted eleven to one in favor of the risky inside option. However, Tom Young, NASA's director of planetary programs, voted one to none for the outside option, and the issue was decided. The particles and fields men wanted to get as close to the planet as possible and damn the torpedoes, but Young was already thinking about the Voyager encounters with Saturn. The deciding factor was the preservation of the Voyager 2 Uranus option. The Uranus aiming point at Saturn was in the region of the suspected E ring; if Pioneer made it through, then presumably Voyager would as well. If Pioneer crashed, Voyager 2 could be retargeted. It was rather as if Lewis had told Clark to check out a cave for grizzly bears.

Following the outside track, the spacecraft would cross the ring plane on September 1 at a distance of 112,000 km from the planet. It would then fly under the rings, passing within 21,400 km of Saturn. Four hours after the first crossing, Pioneer would again pass through the ring plane on the ascending, outbound leg of the encounter. The crossing itself, at 53,000 mph, would last no more than about $\%_{10}$ second.

During the Voyager Jupiter encounters, scientists—no less than the general public—had grown accustomed to the sharp, detailed pictures produced by the Voyager imaging system. But Pioneer was a spacecraft from an earlier generation, and its imaging system was an unfortunate reminder of the way things used to be. It functioned well, but the images were bound to be a disappointment for an audience jaded by Voyager.

If spacecraft were automobiles, Voyager would be a Mercedes and Pioneer a Volkswagen beetle. The entire Pioneer 11 spacecraft weighed just 260 kilograms, versus 825 kg for the Voyager mission module; the Voyager science package weighed in at 117 kg, while Pioneer could carry just 30 kg of scientific instruments. Voyager was descended from the Mariner family, famed for sophisticated, heavy-duty science observations; Pioneer 11, on the other hand, was the last of its line, the final mission in a program begun in 1958, less than a year after the launch of Sputnik.

Pioneer 1 was America's first deep-space probe, sending back data from a distance of 70,700 miles. Pioneer 3 mapped the Van Allen radiation belts around the Earth; Pioneer 4 returned data from beyond the moon; and Pioneer 5 was one of the first spacecraft designed to study the solar wind. Pioneers 6 through 9 were highly successful deep-space probes that monitored the far side of the sun, providing an early warning against large solar flares. Pioneer 6 was designed for a six-month lifespan, but it was still functioning well in 1979, sixteen years after launch.

Unlike the Voyagers, the Pioneer missions were controlled from the Ames Research Center, a NASA facility located at Moffett Field Naval Air Station in Mountain View, California, thirty miles south of San Francisco. Although people from Ames worked on Voyager and JPL personnel contributed to Pioneer, a distinct "us" and "them" attitude was apparent during the Pioneer 11 encounter. Pioneer people were touchy about comparisons with Voyager and insisted that their spacecraft was every bit as good as the flashy competition from Pasadena. The Pioneer imaging system might be inferior, but scientists at Ames honestly believed that their particles and fields experiments were better than Voyager's. Pioneer Project Scientist John Wolfe even claimed that "Voyager didn't find out anything new." The raw data from the Pioneer Jupiter encounters contained everything later discovered by the Voyagers, including evidence of the Jovian ring. No one had noticed it, but the proof was there nevertheless, hidden in the numbers.

Pioneer 11 carried a dozen experiments, almost all of them concerned with particles and fields. The radiation environment of Saturn was of special interest because, unlike Jupiter, Saturn was not known to be a strong radio source. It seemed likely that any Saturnian radiation belts trapped within the still hypothetical magnetosphere would interact with the rings in such a way as to prevent the high-energy synchrotron bursts that were observed from Jupiter. Among the instruments designed to study trapped radiation at Saturn was a Geiger-tube telescope, similar to the instrument aboard Explorer 1 which first detected Earth's Van Allen belts. The principal investigator for the experiment was none other than James Van Allen, for whom the radiation belts were named. Twenty years and three planets later, Van Allen was still counting Geiger clicks from space.

The much-maligned Pioneer imaging system was known as the *i*maging *p*hotopolarimeter (IPP). The system measured light intensity and polarization in addition to making images. By analyzing the light-scattering properties of the Saturnian atmosphere, the IPP could give valuable information on the size and nature of particles in Saturn's upper atmosphere. But the proof of an imaging system is in the pictures it takes, and by that standard the IPP was inevitably disappointing.

The Voyagers are stabilized along all three axes; to make an image, the scan platform simply aims the cameras at the target object. Pioneer was spin-stabilized, meaning that the entire spacecraft rotated 5.7 times every minute around its central axis, which was always pointed back at the Earth. To get a picture, the IPP had to take a strip of images on each rotation of the spacecraft. With a field of view of 0.3 degree, it took between 25 and 110 minutes to assemble a single picture. All told, Pioneer 11 would gather just 150 images at Saturn; the Voyagers collected more than 20,000 at Jupiter.

Pioneer could get color images, but their value was open to some question. Since the system was restricted to red and blue light, the green component of the color images had to be added on Earth. "It's very simple," said Principal Investigator Tom Gehrels of the University of Arizona. "We have two knobs, one is red and one is blue, and you turn them until you think it's about right."

The main limiting factor for Pioneer was the amount of data it could transmit back to Earth. Pioneer's on-board computer was state-of-the-art when it was launched; by 1979 it was an antique. While Voyager would broadcast more than 44 kilobits per second from Saturn, the best Pioneer could manage under ideal conditions was 1,024 bits per second. That was under ideal conditions; problems with terrestrial weather and equipment at the tracking stations would force the mission controllers to switch to a lower data rate of 512 bps during much of the encounter.

The first major event of the encounter would be the crossing of the bowshock at the outer edge of Saturn's magnetosphere. Virtually everyone assumed that Saturn had a magnetic field and a magnetosphere, though no one was very certain about when and where the spacecraft would enter it. James Van Allen said that he expected the bowshock would be found at a distance of about 39 or 40 Saturn radii (R_s) from the planet, but admitted that "this may all be quite wrong." Van Allen described Jupiter's magnetosphere as "very soft and squishy," while Earth's magnetosphere is much more rigid; Saturn's was an unknown quantity, but Van Allen thought it would turn out to be more like Earth's than like Jupiter's.

Whatever the true nature of Saturn's magnetosphere (if it had one), Pioneer would encounter it in abnormal circumstances. A month earlier, Pioneer Venus, still in orbit around the second planet, had observed a large "solar wind event." The normal speed of solar wind particles is about 400 km/sec, but the "event" increased the velocity to 600 km/sec. Even at that speed, it took the particles nearly a month to spiral outward along the sun's magnetic field lines to Saturn—just in time for Pioneer. Meanwhile, a large solar flare had erupted on August 18, and the more energetic flare particles were also arriving at Saturn. The combination of the two solar events, according to Wolfe, had "muddied the waters." Conceivably, the Saturnian bowshock could be as close to the planet as $1/3$ R_s.

One interesting consequence of a small or unusually compressed Saturnian magnetosphere was that Titan might be outside the field. Some theorists thought that Titan's atmosphere might be supplying the magnetosphere with charged particles. If Titan routinely spent part of the time inside the magnetosphere and the rest of the time outside, the field might have some unusual properties. But everyone was still just guessing; a few days before the encounter, Saturn was still an immense riddle.

Titan was the only moon that would be imaged by Pioneer. The other Saturnian satellites were too small and too far away from the spacecraft's trajectory. An image of Iapetus, for example, would have been no more than 9 pixels in diameter, far too small an image to tell us much about the moon's bizarre dark leading and bright trailing hemispheres.

While the wait for the bowshock continued, attention turned to the images. Because of the various problems with the spacecraft and the Deep Space Network, real-time images were unavailable. The raw data had to be transferred to a computer processing facility at the University of Arizona, resulting in a lag of several hours between receipt of the data and the appearance of the actual pictures.

The pictures were worth the wait, if only because they were better than the best Earth-based photos. However, Saturn itself was something of a disappointment. Ringworld was turning out to be Blandworld. The golden cloud tops of Saturn were nearly featureless, and the horizontal belts and zones that were so prominent on Jupiter were few and faint. A few small atmospheric spots were present, and these would aid in the determination of Saturn's precise period of rotation. But the multichromatic cornucopia of swirls, blobs, jets, and whirlpools seen on Jupiter was not in evidence at the sixth planet. However, the experience of the Pioneers and Voyagers at Jupiter argued against jumping to conclusions. The Pioneers had also seen a bland Jupiter in 1973 and 1974, while the Voyagers found an utterly different Jupiter in 1979.

The unexpected discovery of the Jovian ring had given scientists a sneak preview of Saturn's much more spectacular ring system. The extremely small size of Jupiter's ring particles—evident in the way they "forward-scattered" sunlight—was a hint that Saturn's rings might turn out to be more complex than anticipated. Over the years, many theories had been put forth to explain the rings, but it is probably fair to say that, by the eve of the first Saturn encounter, no one had a great deal of confidence in any of them. Rings might have been formed from unconsolidated junk, from a shattered moon, or from material sputtered off the surfaces of nearby satellites—or from some as yet unsuspected process. The ring particles were almost certainly composed of water ice and rock but in what proportion no one could say. The overall structure of Saturn's rings seemed to be determined by gravitational resonances with satellites, such as the one which apparently formed the Cassini Division. But all the ideas about rings and their nature had been developed from a billion miles away. The first close look at the rings was bound to produce surprises.

Pioneer's view of the rings looked totally unlike the familiar Earth-based pictures. Instead of broad, bright rings, the images showed alternating bright and dark bands. Even experienced astronomers found the images confusing on first viewing. The sun was about 2° below the plane of the rings, so that sunlight fell on the bottom, or southern side, of the rings. Approaching from above the rings, the spacecraft had a never-before-seen view of the unilluminated side of the rings. In effect, the Pioneer images were like photographic negatives, reversing the normal patterns of light and dark.

Where the rings appeared bright, sunlight was leaking through gaps or thin places; where they appeared to be dark, the rings were thick enough to be opaque. However, the dark regions could also be indicative of a total lack of ring material; with no small particles present to reflect sunlight, an empty region would look the same as a dense region. This seemed to be the case with the suspected innermost D ring, which was completely dark. Either the D ring was unexpectedly dense, or it simply didn't exist. The outermost (and unconfirmed) E ring was another area of ambiguous blackness. On August 30, Gehrels reported that scans of the region outside the A ring were being conducted in the hope of finding evidence of the E ring's existence; "So far," he said, "nothing doing."

The size and distribution of the ring particles would determine whether or not Pioneer survived the ring-plane crossing. There were many theoretical calculations about the population of ring particles, but no one actually knew how numerous or how big they were. If the particles were big—on a scale of meters to tens of meters—then it seemed unlikely that the spacecraft would hit anything. If the particles were extremely small—micron (a millionth of a meter) size—then collisions were inevitable but probably unimportant. But if the typical size of the particles was in the range of millimeters to centimeters, crossing the ring plane could be like flying through a cloud of buckshot.

With the ring-plane crossing in mind, the blackness from the region of the E ring was encouraging. It implied either large (and thus, widely spaced) particles or a general absence of particles of any size. Killer-size particles ought to have reflected at least some light if they existed. Since the only evidence for the E ring came from photos taken at the Allegheny Observatory in 1966 (the last time Earth was favorably positioned to search for new rings), some Pioneer scientists were beginning to have doubts about its reality.

Looking over the shoulders of the Pioneer investigators was a small band of interested Voyager scientists who had come to Ames for the encounter. Aside from scientific curiosity, the Voyager scientists had been drawn to Ames by the hope of getting a sneak preview of their own encounter with the sixth planet. Pioneer would help establish more precise positions for some of the moons and rings, providing vital information on where to aim Voyager's cameras and other instruments.

Unlike JPL where the scientists and press each have their own buildings, at Ames everyone was crowded into the same building. The auditorium for press briefings was small, overflowing, and impossibly hot when the television lights were on. The close quarters, however, lent an air of intimacy to the encounter and a nice Northern California mellowness that contrasted with the hectic hustle of Los Angeles.

Early on the morning of August 31, Pioneer at last found the Saturnian bowshock. John Wolfe told reporters that he had made a bet on the bowshock crossing with James Van Allen. Van Allen won, and Wolfe ceremoniously presented him with a check for fifty cents. Van Allen looked at it dubiously and said, "I'd be much happier if this were certified."

Not even the bowshock crossing was certified. Soon after first contact with the Saturnian magnetic field, the solar wind pressure shot upward and Pioneer was back in interplanetary space. The first time this had happened, during the Pioneer 10 Jupiter encounter, Wolfe came into the press area at 2 AM and confessed that he had no idea what was going on. But now, a pulsating magnetosphere was a familiar phenomenon. Pioneer 11 first crossed the bowshock at about 24 R_s and stayed inside it for half an hour. After playing tag with the bowshock for most of the day, by 5 PM the spacecraft was inside the magnetosphere to stay, with the final crossing coming at a distance of just 17 R_s from the planet.

The big news in this, of course, was that Saturn definitely had a magnetic field. It joined Mercury, Earth, and Jupiter as the only planets known to possess a field. But the Saturnian magnetic field was an odd beast, and its discovery had a ripple effect that shook the foundations of the basic theory of planetary magnetism.

The strength of the Saturnian magnetic field was about what had been expected. It was measured to be roughly 500 times stronger than that of the Earth and about 35 times weaker than the Jovian field. At the Saturnian cloud tops, the field's strength was only about $\frac{1}{3}$ Gauss, or close to the strength of the terrestrial field at the surface of the Earth. The recent solar wind events had compressed the field to a smaller than normal size, but John Simpson noted that even under normal conditions Saturn has "a modest-sized magnetosphere, compared to the planet."

The truly baffling thing about Saturn's magnetic field was its near-perfect alignment with the planet's axis of rotation. On other planets and on the sun, the magnetic poles are offset from the rotational poles by about ten degrees; on Saturn, the offset was less than one degree. Theorists believed that the divergence between magnetic and rotational axes was important to the creation of the internal dynamos that produced magnetic fields in the first place. But Saturn, not for the last time, stood in silent defiance of the theorists. Pioneer had barely arrived at Saturn, and already it was upsetting long-held beliefs.

Pioneer 11's moment of truth came at 9:02 AM Pacific Daylight Time, on September 1. At that instant, the spacecraft passed through the plane of the rings. Eighty-six light-minutes away, on Earth, anxious scientists knew that their spacecraft had already survived or been destroyed; there was nothing for them to do but wait for the signals to arrive. The stream of data from Pioneer was like a billion-mile-long pearl necklace, with 5.5 million pearls strung between Saturn and Earth. The pearls were spaced 182 miles apart—that is, traveling at the speed of light, each transmitted bit of data moved 182 miles before the next bit was broadcast. There would be no time for a warning if the spacecraft collided with something; the stream of pearls would simple stop.

As the Ames clocks slowly ticked off the minutes until 10:28, the clock watchers lit up cigarettes, chewed on fingernails, and joked nervously. At stake that morning was nothing less than the future of outer solar system exploration. If Pioneer failed to make it through the ring plane, it might prove impossible to design a safe trajectory for Voyager 2 that would still permit a flight to Uranus: and if Voyager 2 did not encounter Uranus and Neptune, it seemed certain that mankind would not get a close look at those outer giants until well into the twenty-first century.

The appointed moment arrived, and the signals from the spacecraft continued unabated. A minute passed, then another, but there was still no word from Project Manager Charles Hall. Finally, eight minutes after the predicted moment of ring-plane crossing, Hall said that he thought they had made it; Pioneer 11 was still alive.

There were obligatory cheers, but mainly, there was a deep sense of relief. Dave Morrison, of the Voyager imaging team, grinned and said, "On to Uranus!"

Pioneer dipped through the ring plane at a distance of 2.82 R_s from the planet, some 38,000 km outside the visible edge of the A ring. Speeding along just below the rings, the spacecraft closed to within 21,000 km of Saturn's golden cloud tops, accelerated by Saturn's gravity to a maximum speed of nearly 70,000 mph. Saturn's gravitational field bent the probe's trajectory, sending it halfway around the planet before it ascended back through the ring plane at a mirror-image distance of 2.82 R_s. During its high-speed passage under the rings, Pioneer was moving too fast for the imaging system to be able to take any pictures, but the particles and fields experiments more than made up for the absence of images.

The spacecraft carried a micrometeoroid detector which

counted impacts by micron-size particles. The instrument had been designed to take data during the months-long crossing of the asteroid belt on the way to Jupiter and was of limited value in counting impacts during the brief ring-plane crossing. The instrument consisted of 234 gas-filled cells that released their gas when punctured by an impact; unfortunately, after a cell was hit, the instrument took 80 minutes to recycle for further readings. The detector recorded a hit some 10 minutes before ring-plane crossing, when the spacecraft was still more than 1,000 km above the rings. There should have been few, if any, particles in that region, and to some scientists the impact implied that the rings were unexpectedly thick. There was no way to tell how many more hits the spacecraft might have taken during the crossing, but Charlie Hall commented that "we may have been lucky."

Pioneer passed through the region of the E ring unscathed, but that didn't mean that the ring wasn't there. Despite the lack of imaging, the E ring showed up clearly in the particles and fields data. Van Allen and his colleagues were using their instruments in a new and creative manner, a technique Van Allen called "particle beam astronomy."

The Pioneer experimenters measured the intensity and population of protons and electrons as they spiraled around the magnetic field lines in the magnetosphere. Any material—such as rings and satellites—that intersected the lines of force would absorb or block the protons and electrons, reducing the overall particle count. One such dip in the particle count occurred in the area of the suspected E ring, at the orbital distance of the moon Enceladus. The ring was apparently too diffuse to show up in the images, but the particle data implied that it existed nevertheless.

The effect may have been subtle at the diffuse E ring, but at the edge of the visible rings, 2.33 R_s from the planet, the particle dropoff was unmistakable. Van Allen called it "the dramatic guillotine effect." As Pioneer passed under the rings, the particle count suddenly dropped to almost nothing. "This is the lowest count we've had since we were sitting on the launch pad at Cape Canaveral," said Van Allen. The rings constituted "an opaque umbrella," which made the near-ring environment "the best shielded place in the solar system." Other experimenters saw the same effects in their data; John

Simpson, PI for the charged-particle composition experiment, said his results looked "as if they took our instrument off the spacecraft." The extinction of charged particles in the ring shadow was nearly complete. Even under the broad Cassini Division, the count remained close to zero. This was the first hint that the supposedly empty Cassini Division was not so empty, after all.

On the afternoon of the 1st, while Pioneer was still cruising beneath the rings, Imaging Team Leader Tom Gehrels announced three significant new discoveries. The first was the acquisition of an image of the north polar region of Saturn. Amid the golden hues of the atmosphere, scientists found definite evidence of Jovian-style belts, zones, whirls, blobs, and scallops. A high-speed jet stream apparently existed near 70°N. Saturn had its own distinctive style, but it was clearly a close relative of Jupiter.

Next, Gehrels announced what may have been the discovery of a new satellite. At about 2.5 R_s, Pioneer's cameras spotted a small object about 200 km in diameter. It may have been a brand-new moon, or it may have been the elusive moon Janus. Janus was discovered from Earth in 1966, but later observations found that the moon was in the wrong place. No one was certain whether Janus was one moon or two (or none), and the Pioneer image only seemed to compound the confusion.

Concerning Gehrels' third announcement, there was no ambiguity. Pioneer had discovered a new ring, the F ring, just 3,000 km beyond the visible edge of the bright A ring. The F ring was barely 500 km across, a very un-Saturnian structure. In its narrowness, it resembled Jupiter's newly discovered ring or the nine slender rings around Uranus. The F ring also showed up in the particles and fields data.

Gehrels and his colleagues immediately christened the 3,000-km gulf between the A and F rings the "Pioneer Gap." Although it seemed to be an inspired choice of nomenclature, resonating with memories of the Cumberland Gap and the travels of earlier American pioneers, the name didn't survive for long. It quickly became the "Pioneer Division," a name more in keeping with the existing labels of the Cassini and Encke Divisions. By the time Voyager 1 reached Saturn, even "Pioneer Division" had been discarded. Voyager scientists were less than charitable about the Pioneer nomenclature.

PIONEERING

"These names have not been accepted by the IAU,*" said Brad Smith, "and we are not using them."

The next day, September 2, scientists trooped before the press in two platoons for "quick-look results day." Andy Ingersoll, principal investigator for the infrared radiometer experiment, reported, "Our instrument yesterday measured the temperature of everything in sight." The infrared readings revealed that the rings measured about 70°K at a low sun angle and 64°K in the shadows. The instrument saw different temperatures in the Saturnian belts and zones, implying a complex heat flow within the atmosphere. The overall average temperature of the clouds was −180°C, slightly warmer than had been expected. The result implied that Saturn was radiating from two to two and a half times as much heat as it received from the sun. This was too much for the primordial heat model that applied to Jupiter. Something else was going on at Saturn, and the separation of helium in the atmosphere seemed to be the logical candidate. The existence of this internal "helium rain" implied that Saturn's internal structure had to be significantly different from that of Jupiter.

Saturn was also challenging at ultraviolet wavelengths. Pioneer's ultraviolet photometer found evidence of polar auroras, similar to the auroras observed on Jupiter. There was also a distinct emission from the region of the B ring. Some scientists believed that this implied a very diffuse "atmosphere" of water vapor and hydrogen hovering above the rings. Charged particles colliding with the "atmospheric" molecules would supply the energy for the emissions. The ultraviolet detector also found a dim, diffuse glow originating in the region near the orbit of Titan, which may have been evidence of an orbiting cloud of gas or plasma—a neutral hydrogen torus—that was produced by charged-particle interactions with Titan's upper atmosphere.

Titan, of course, was the object of intense speculation, but Pioneer would provide no definitive answers to the many enduring questions about this unique satellite. The best image that was obtained of the moon was just 35 pixels across and revealed almost no detail; there was a hint of banding in the Titan cloud tops, though to

---

* International Astronomical Union. This group meets every three years and, among other things, assigns official names to newly discovered objects and surface features.

the untrained eye the images looked like nothing more than a fuzzy, featureless orange ball.

The Pioneer scientists felt fortunate in getting even a few unrevealing images of Titan. Uncertainties about Titan's orbit and the spacecraft's trajectory made it difficult to aim the cameras. Additional complications were created by the switch from a 1,024 bits per second data rate to 512 bps. Bad weather at the tracking stations necessitated the reduction, which made it hard for the imaging team scientists to know exactly when the shutters should be opened during a given spin of the spacecraft. All they could do, said Martin Tomasko of the University of Arizona, was "point the telescope and hope for the best." Their best wasn't at all bad, though the first Titan image clipped off the top of the moon. Some rapid recalculations got the satellite centered in subsequent images.

The measurement many scientists wanted most was Pioneer's temperature reading of Titan's atmosphere. Atmospheric temperature was the key to many of the competing theories about Titan, and it was hoped that Pioneer would provide enough data to put tighter constraints on the various hypotheses concerning the thickness of Titan's atmosphere. But on the morning of September 3, Project Manager Hall reported that the Titan infrared data had been lost. It rained in Spain again, and in Australia too, forcing another drop in the data rate, from 512 to 256 bps. It seemed that the playback of the Titan data had been garbled by thunderstorm interference, and Hall was doubtful that much of the signal could be sorted out from the noise.

By the next morning, however, the Titan story took a bizarre turn. It now seemed that the temperature data had been lost due to the inadvertent interference of a Soviet satellite. The Russian Cosmos Earth satellite broadcast in the same frequency range (2,290 to 2,300 MHz) as Pioneer and apparently drowned out the signal from Saturn. Overlapping frequencies were nothing new, and NASA, via the State Department, routinely secured Soviet cooperation in such matters. But this time, possibly because of the long Labor Day weekend, someone somewhere forgot to ask the Soviets for their cooperation. In order for the receivers to be able to accept a signal from Pioneer, they had to pick up the eight-bit address code at the beginning of a message; apparently, the signal from the Soviet satellite—a

From a range of 8 million miles, Voyager 2 caught Io, innermost of the Galilean satellites, passing in front of Jupiter's southern hemisphere.

This Voyager 1 image reveals the incredible activity surrounding Jupiter's Great Red Spot. From 3 million miles away, the smallest features visible are some 55 miles across.

A half-dozen Earths would fit into the vortex of the Great Red Spot. Its counterclockwise rotation produces turbulence in the atmospheric flow patterns, as seen in this Voyager 2 image.

The surface of Europa, smallest of the Galilean satellites, resembles a cracked eggshell. The icy surface may cover a subsurface ocean of liquid water. The bright white streak just below center may consist of fresh ice being pushed up from beneath. Such a resurfacing process could account for the nearly total absence of impact craters. This Voyager 2 image is from a range of 150,600 miles.

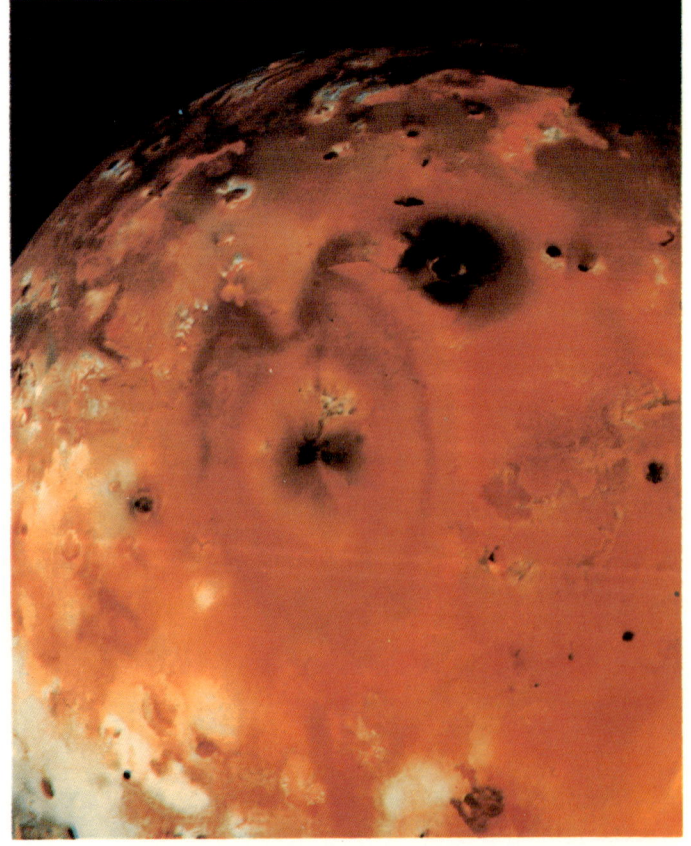

Io, seething with lava, glows in brilliant red, yellow, and orange. Sulfur, which takes on different hues at different temperatures, is responsible for its colorful surface. Large black spots are sulfur lakes. Scientists recognized fuzzy black spots as active volcanoes. This stunning image was taken by Voyager 1 from a distance of 226,200 miles.

Ganymede, largest moon in the solar system, revealed its crazy-quilt surface in this Voyager 1 image from 1.6 million miles. Composed of an ice/rock mixture, Ganymede's surface has been subjected to large-scale tectonic faulting during its 4.5-billion-year history.

This false color Voyager 2 image reveals Callisto's violent past. The heavily cratered ancient surface resembles terrains on Mercury and Earth's moon. Bright spots consist of fresh ice ejected by relatively recent impacts.

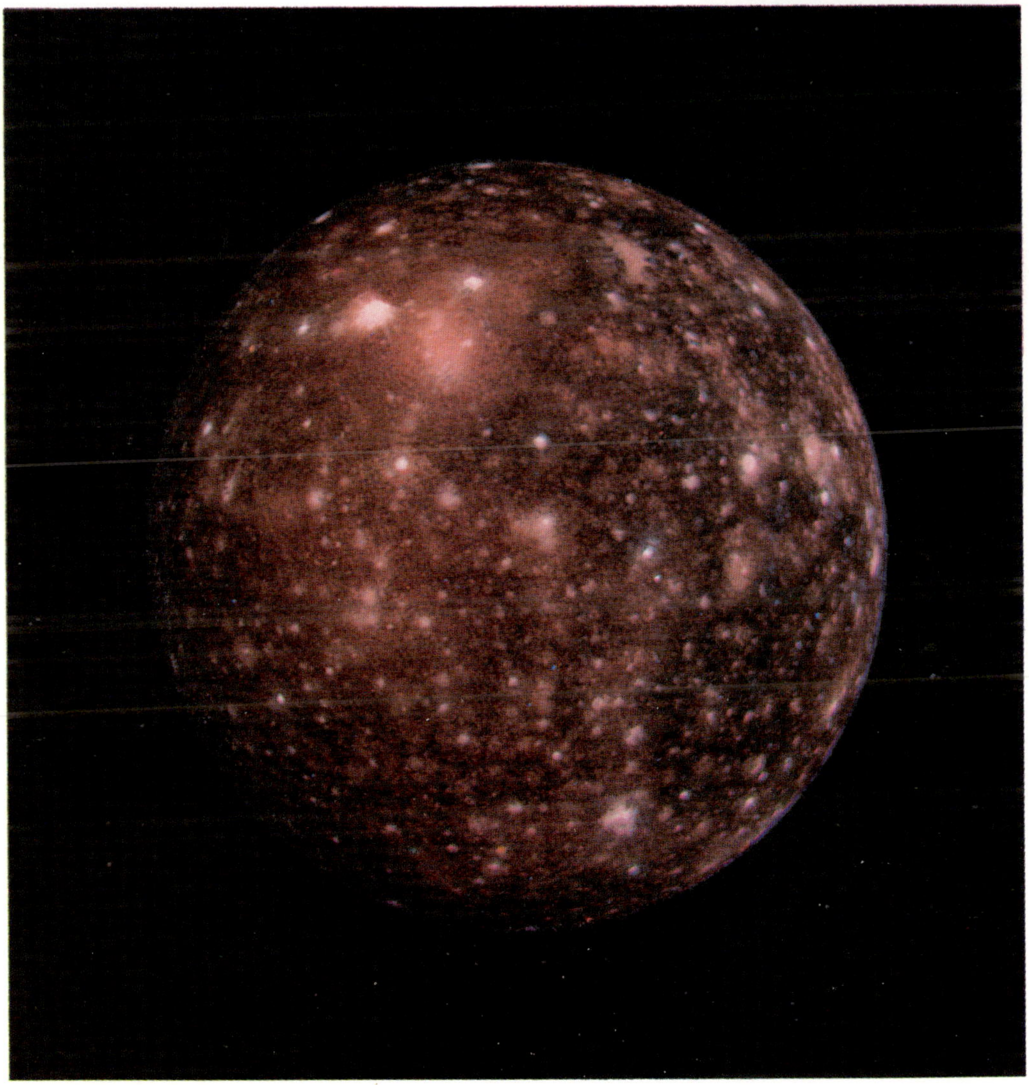

Taken 9 months apart, these Voyager 1 and Voyager 2 images show subtle changes in Saturn's muted atmosphere. Moons Tethys (above) and Dione are seen in the earlier image.

This computer-enhanced Voyager 1 image reveals a view never seen before—Saturn and its rings from the night side.

From 393,000 miles, atmospheric flow patterns near Saturn's north pole are visible in a Voyager 2 image. Bright spots may be tops of convective cloud systems. The smallest visible features are about 10 miles across.

A false color Voyager 2 image reveals the profusion of ringlets in Saturn's C-ring and (top left) B-ring. The colors are computer-manipulated to indicate differing size and composition of B- and C-ring particles. Scientists, expecting each ring to contain particles of a characteristic size, were surprised to note the presence of "yellow" B-ring material within the "blue" C-ring.

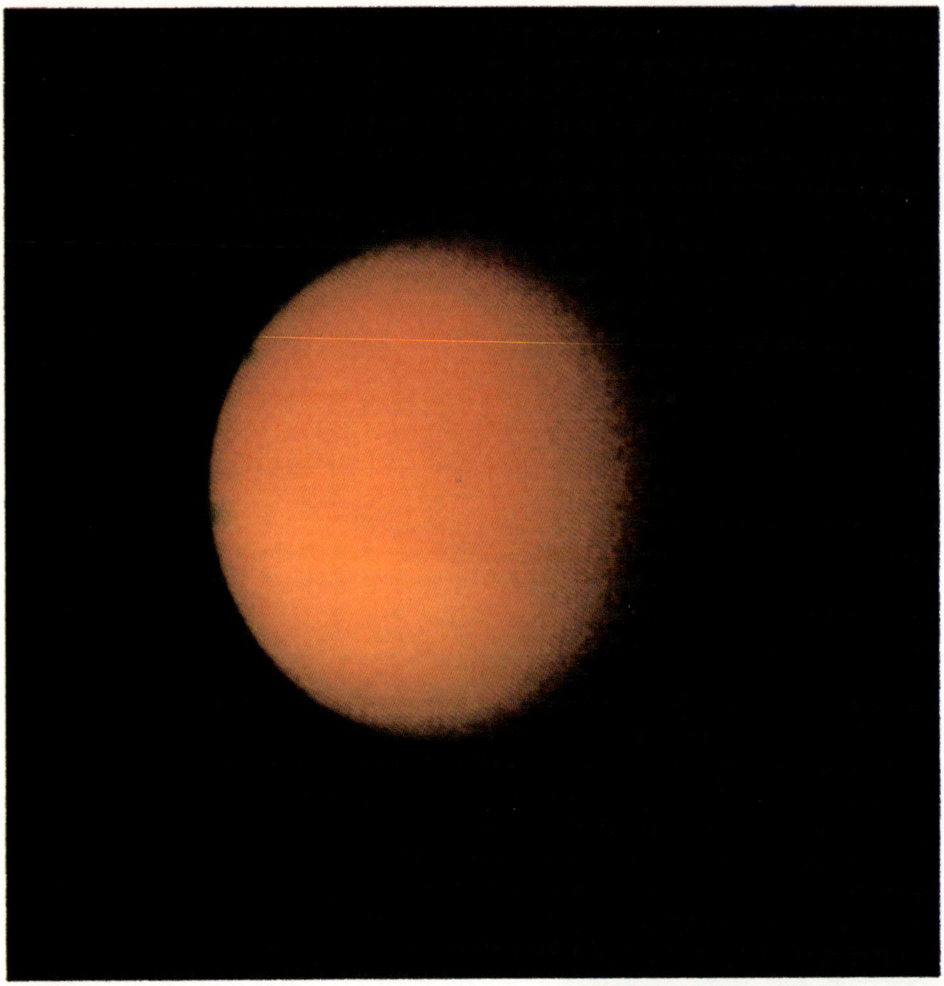

These Voyager 2 images of Saturn's moon Titan show the dense, almost featureless atmosphere of the solar system's second-largest moon. The atmosphere is denser than Earth's and scatters sunlight to produce a halo effect when seen from the night side. The only moon known to possess an atmosphere, Titan's orangish clouds are composed of nitrogen and methane. The clouds conceal a surface which may feature seas of liquid methane and a kilometer-thick layer of organic debris produced by chemical reactions in the atmosphere.

Saturn's tiny (240-mile diameter) moon Mimas was subjected to early bombardment which may have come close to shattering it. This Voyager 2 image from 80,000 miles reveals features as small as 1 mile across.

A mosaic of Voyager 2 images from 74,000 miles displays the complex surface geology of Saturn's moon Enceladus. Long, straight lines are faults. A smooth area of young terrain bisects an ancient, heavily cratered region. Complete absence of impacts in the young terrain suggests that Enceladus is still geologically active.

A color mosaic of Voyager 1 images shows the battered, ancient surface of Saturn's moon Dione. The largest craters visible are about 60 miles across. Bright rays on the left limb of the moon are probably composed of ice ejected by impacts. The sinuous valley at right may be the result of faulting in Dione's icy crust. Images were taken from a range of 100,000 miles.

Saturn's smaller moons such as Mimas may have narrowly escaped the fate of Hyperion, seen in a Voyager 2 image from 300,000 miles. This irregular moon (150 by 225 miles) was probably formed by large, devastating impacts on the original body. Unlike all other moons, Hyperion may not be in rigid tidal lock with Saturn: impacts and the influence of Titan may have distorted its original orbit and orientation.

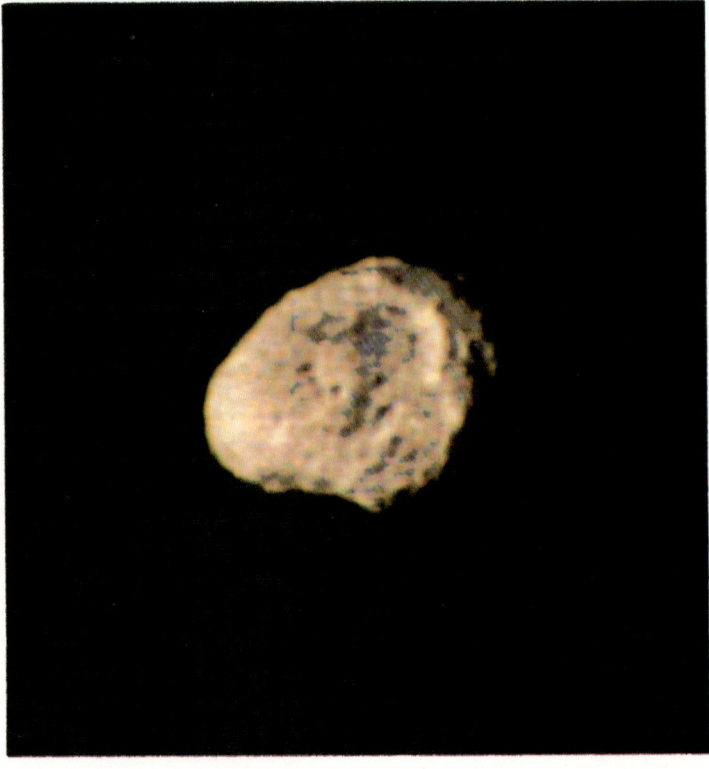

hundred to a thousand times stronger than the Pioneer broadcast—obscured the vital code and the data was lost.

Ironically, the problem of too many broadcasts on not enough frequencies was then under consideration at the World Administrative Radio Conference in Geneva. Radio frequencies are a limited natural resource, and the sudden profusion of Third World radio stations, communications satellites, and CB radio rigs had seriously overloaded the radio spectrum. Space scientists were trying to get a frequency reserved for deep-space probes; in the meantime, interference was a continuing problem. Low-flying planes often disrupted the reception at the Goldstone antenna; and during the Pioneer Venus mission, the Chilean Army had blocked off the Pan Am Highway to prevent CB interference with NASA's South American receiver. But this time, it seemed, the problem was simply a bureaucratic fumble. NASA put out a carefully worded press release that somehow managed to exonerate the Soviets without fixing the blame on anyone else.

On September 6, the situation unscrambled itself, leaving the participants in the affair looking just a bit foolish. Miraculously, the Titan infrared data had been received, after all. Andy Ingersoll explained what had happened. The night the data came in, a weary Ingersoll watched the monitors as the string of numbers arrived from Saturn. No one was sure exactly when the temperature data would come in, and there were so many data drops because of the bad weather that no one realized it when the address numbers were picked up. Ingersoll flew back home to Caltech for the weekend, thinking the data was irretrievably gone. Later, working with JPL technicians, Ingersoll looked at the numbers again, "and sure enough, there was Titan—sprinkled in among data dropouts, which appear as dollar signs on my printout."

The Soviet satellite transmission had ended moments before the temperature readings of Titan's atmosphere came in, and the combination of thunderstorms, scrambled printouts, fatigue, the holiday weekend, and "a lot of little things" led to the presumption of Soviet interference. As it turned out, very little data was lost, and none at all due to the Soviet satellite. The whole thing, Ingersoll said, "was marginally my fault, but I don't even feel bad about it."

The infrared data was barely worth the trouble it had caused. Ingersoll reported that the temperature at the cloud tops of Titan was 75°K ±5—"about what you'd expect." The temperature seemed to be uniform across the globe, but there were so few data and so much noise that "there's no prayer of measuring a day–night temperature difference in these data." Titan was still the mystery it had always been.

In the midst of the confusion over the temperature data, John Wolfe announced an important and completely unexpected discovery. Twenty-three minutes after crossing the ring plane on the inbound leg every one of the spacecraft's particle detectors registered a sharp and sudden drop to the background level—"as if we went by a very large rock." Something had gotten in the way of the spiraling charged particles, completely blotting them out—the guillotine effect again. "We would have to have been pretty close to something pretty large," said Wolfe. "It's very intriguing."

The unseen object became known as the "Pioneer Rock," another good new name that didn't last long. The "rock" was found at 2.53 R_s as the spacecraft was zooming along just 2,000 km below the ring plane. Since the object was almost certainly orbiting in that plane, Pioneer apparently had a very close brush with disaster. Analysis of the data indicated that the object was about 100 to 300 km in diameter, and the spacecraft had flown right through its magnetic "wake."

Officially, the Pioneer Rock became 1979S2. The tiny new satellite (1979S1) imaged earlier also orbited Saturn at a distance of 2.53 R_s, and some quick calculations showed that S1 and S2 were very likely the same object—although neither of them seemed to be the notorious Janus, whose existence was still unconfirmed from both Earth and space. Whatever the satellite's identity, the unexpected close encounter with it had been hair-raising. "We were damned lucky," commented one scientist.

The technique of "particle beam astronomy" also led to the discovery of yet another ring—much to the annoyance of journalists who had been amusing themselves by devising new mnemonic phrases which described the sequence of rings, D–C–B–A–F–E. Gems such as "Disco Can Benumb A Fine Ear" and "Dense Clouds: Boastful Astronomers' Favorite Excuse" were rendered obsolete by the discovery of a G ring outside the known rings. James H. Trainor, a cosmic ray experimenter from NASA Goddard, told the press on September 6 that "something's eating up the protons" at an orbital distance of between 10 and 15 R_s. The presentation of the data was so low-key that journalists had to ask Trainor if he had, in fact, just announced the discovery of a G ring. Trainor assured them that he had.

Imaging Team Leader Tom Gehrels clarified matters by recapitulating Pioneer's adventures with the rings. The innermost D ring, supposedly seen from Earth (by Brad Smith, among others), was nowhere to be found. Similarly, there was no evidence for the Guerin or French Division, which separated the Cheshire cat-like D ring from the C ring. The C ring was seen clearly, and the B ring turned out to be denser than expected. The Cassini Division seemed to have quite a lot of material in it, another unexpected discovery; at one time, there had been talk of sending Pioneer right through the Cassini Division, a course that almost certainly would have led to disaster.

In the A ring, the spacecraft found considerable detail. Beyond it, the Pioneer Gap was another new discovery, as was the "exquisitely fine feature" of the F ring. The E ring, if it truly existed (particles and fields data implied that it did), was too diffuse to be imaged by Pioneer. And now, the G ring joined the list of unseen but undoubtedly real structures orbiting Saturn.

The Saturn encounter was over. Already, the data rate had been reduced to a miniscule 64 bits per second, and by the evening of the 6th it would be cut down to 32 bps. Saturn was getting so close to the sun, as seen from Earth, that any sort of communication with the spacecraft was increasingly difficult, due to solar interference.

In the long run, Wolfe was hopeful that Pioneer 11 would keep broadcasting outer solar system data for years. Facetiously, he explained, "The DSN has been able to improve its capabilities faster than Pioneer has been leaving the solar system. If you extrapolate that out, we can track Pioneer forever." In truth, "forever"* would

---

* "Forever" may come even sooner than 1987. In another budget-cutting move, the Reagan Administration decided to turn off all the surviving Pioneer spacecraft, including Pioneer 11, on January 1, 1983. See Chapter 10.

come by about 1987, when Pioneer's nuclear generator would be exhausted. Still, the remaining eight years would see Pioneer reaching a distance of some 35 to 40 AU, putting it some three and a half billion miles from the sun by the time its power supply expired. As it traversed this unexplored region of the solar system, the spacecraft would carry on the task it had performed so well for so long—pioneering.

9

NO SMALL RAPTURE

A cold, blustery wind swept down out of the mountains, and through the city of Tucson, making the planetary scientists gathered there feel right at home. Space scientists had been facing the chill winds of professional extinction for years, but their future had never looked quite as bleak as it did during their October, 1980, convention in Tucson. Voyager 1 was less than a month away from Saturn, but even that exciting prospect did little to lift spirits.

Scientists and the handful of journalists who came to the four-day meeting of the Division of Planetary Sciences of the American Astronomical Society arrived at the Tucson Ramada Inn expecting to learn about the planets; the first thing most of them learned was the shocking news of Tim Mutch's death in the Himalayas. NASA's associate administrator for space sciences, Mutch was universally liked and respected. He was widely regarded as the most effective advocate for planetary exploration within the upper echelons of NASA.

Compounding the shock was the news that NASA Administrator Robert Frosch intended to resign his post. Several other high NASA officials were rumored to be on the verge of transfer or departure. The 1980 presidential election was bound to bring still more changes.

The changes did not bode well for the American planetary exploration program; on the eve of what was to be one of its greatest triumphs, it was on the brink of becoming a thing of the past, rather than of the future. Since 1962, when Mariner 2 flew past Venus, there had not been a twelve-month period without a planetary encounter by an American spacecraft. But now, there remained only the two Voyager flybys of Saturn. Plans for new missions died stillborn in the offices of NASA and JPL. Existing programs, such as the Galileo mission to probe the atmosphere of Jupiter and *V*enus *o*rbit*i*ng *i*maging *r*adar (VOIR), were in trouble and had been postponed until the mid- and late eighties. Most depressing of all, perhaps, was the fact that with each passing day, the once-in-a-lifetime chance for a mission to Halley's Comet was slipping away. The only certain planetary encounter in the eighties would be the Voyager 2 flyby of Uranus in January, 1986, and even that depended on the continued survival of an already ailing spacecraft. In the meantime, Mercury, Venus, the Earth's moon, Mars, Jupiter, and Saturn, worlds we had come to know in some detail, would resume their four-billion-year silence.

The science of space had drawn the scientists to Tucson, but the politics of space dominated the conversations in the hallways and lobbies. Belatedly, the planetary scientists were attempting to get organized in some meaningful and effective way that could influence the decision-making process in NASA and Congress. As a government agency, NASA itself was not supposed to be involved in politics. For more than a decade, NASA had been staggering under the immense weight of the Space Shuttle, and it seemed to some observers that the space agency had been taken for a ride by the Pentagon. The Air Force wanted the shuttle, but it didn't particularly want to pay the $15 billion price tag from its own budget. So NASA ended up picking up the check for the shuttle feast, only to find that what remained in its pocket was little more than cab fare. Space science and planetary exploration were left to dine on crumbs. Now, after years of quietly behaving like Oliver Twist asking for more gruel, the planetary scientists were trying to speak with a strong, unified voice.

The problem was that no one was sure to whom that unified voice should be speaking. The election of Ronald Reagan seemed probable, but he had yet to say a single word on the subject of space exploration. Optimists in the space community noted that Reagan's conservative backers had traditionally favored the development of new space technology, although their record on planetary exploration was ambiguous. Pessimists pointed out that Reagan seemed misinformed about or disinterested in science and could not be relied upon to support research.

Scientists who confidently blazed trails into the cosmos seemed less certain about what course to chart through the political murk of Washington. "If you look forward," said JPL Director Bruce Murray, "there is very, very little . . . we're marching backward." Morale within NASA was terrible, and the cumulative effect of so many cancelled missions and blunted hopes was visible in the exodus of many of the space program's best people. "It's not obvious how you hold together teams of people," lamented Murray. "You have to have a shared, believable dream—and that dream has eroded."

Yet even Murray, a veteran of many funding battles, seemed unsure about what should be done. He warned that while scientific advocacy of specific missions and programs was important and useful, the scientists should avoid overselling their own importance. "Scientists must be partners in a national effort," he declared. "We must not regard ourselves as being privileged, special beneficiaries." What was needed most, to hold everything together, was "a common sense of the history and value of what we're doing—that's the glue."

The scientists had heard all this before, of course—they'd said it themselves often enough—but they had to wonder if anyone else was listening. At least one person was. The guest speaker at the DPS banquet was novelist James Michener, who was considering writing one of his long, best-selling (inevitably) novels about the exploration of space. The scientists loved the idea and urged Michener to follow through on the project. Michener was precisely the sort of ally the space program needed if it was to survive.

Michener told the scientists, to their pride and delight, that seeing the Mariner 9 photos from Mars was "the highlight of my life, intellectually. . . . It blew my mind! I figured, if men could do that, they could do anything." Michener, ever conscious of history, saw the era of space exploration as a major milestone in mankind's development: "We have entered an age of dazzling intellectual brilliance. We are in an age which is the exact equivalent of the Copernican age, when the basic concepts are being extended. . . . I have an absolute conviction that the consequences of this [space exploration] are comparable." But Michener, too, was at a loss to explain why people weren't more aware of this magnificent enterprise. "Why does not the public realize it?" he asked. "I stand here puzzled."

Michener had no answers, no more than Murray, and even his inspiring speech could not dispel the gloom that hovered over the meeting. At best, Michener provided some encouragement to a group that needed it badly. "In a kind of Dark Age," he said, "we should keep our candles burning. . . . I think it is absolutely vital that we preserve the cadre, that we keep the team together." To Michener, the exploration of space was directly linked to the quality of life on Earth: "I think that maintaining a sense of wonder is absolutely essential to a good life," The scientists, who were in the wonder business, gave him a rousing ovation.

After the applause had died down, however, the basic problem remained. With few exceptions, scientists prefer to avoid the bare-knuckle political infighting apparently so necessary in Washington, and when they do plunge into the fray they are often defeated by their own political naiveté. There was a strong sense of common purpose at Tucson, but little sign of a common strategy for dealing with the threat to planetary exploration. To many of the scientists, the value of the planetary program was self-evident and didn't need to be defended in a political street fight. At a time when the scientists desperately needed someone to play the tacky and thankless role of a Machiavelli, many of them seemed content to go on playing the part of Joan of Arc—doomed but pure.

It was in this atmosphere of gloom and desperate hope that the scientists met to consider their latest findings. If one could keep one's mind off of politics, there were plenty of reasons to be excited by what was happening in the planetary sciences. The first detailed surface maps of Venus and the Galilean satellites were now available; there was new and challenging evidence for the existence of wet Martian oases; and a few brave souls even presented papers on Saturn and its moons. The Saturn papers were offered cautiously; when pressed for details, one scientist replied simply, "Just wait another month."

But the Earth-based astronomers had not waited. During the year since the Pioneer flyby, astronomers had been paying close attention to Saturn. Saturn's slow nodding motion had brought the rings back into the plane of the ecliptic for the first time since 1966. Seen edge-on, the light from the rings was minimal, a circumstance which made it possible to conduct telescopic searches for small inner satellites and dim outer rings. The results were both spectacular and mystifying.

Early in February, 1980, Brad Smith and his colleagues at the University of Arizona's observatory spotted what seemed to be a new Saturnian satellite, provisionally named 1980S1. But the new body had a rotational period of 16.67 hours, which put it in the same orbit as the perplexing Janus-objects seen in 1966 *and* the Pioneer Rock (1979S2). A month later, at Mauna Kea Observatory, Dale Cruikshank found yet another new moon, 1980S3, in the same orbit. S3 was definitely a separate body from S1; this was the first time astronomers had ever found two distinct objects sharing the same orbit. It was impossible to be certain which was which, however, and the two new moons (officially, S-10 and S-11) were referred to interchangeably as the co-orbitals.*

At almost the same moment, French astronomers at the Pic-du-Midi Observatory discovered *another* co-orbital moon. This one, 1980S6 (officially, S-12), shared the orbit of Dione. It apparently oscillated around one of the stable Lagrangian points in Dione's

---

*Each new object seen is assigned a temporary designation, in order of discovery. The official, permanent designations are not given until the precise identities of the bodies can be determined. Thus, the same moon might be "discovered" three or four times, with each sighting producing another temporary number, all of which are discarded when the permanent designation is assigned.

orbit, some 60 degrees ahead of Dione.* This new satellite was also known as Dione B, and by now the confusion of new names and designations was bordering on the absurd. Even the astronomers who discovered them had trouble keeping track of which moon was where. The Voyager encounter would soon compound the confusion.

Meanwhile, William Baum, working at the Lowell Observatory, found the first indisputable visual evidence for the existence of the nebulous E ring. This relatively narrow ring hugged the orbit of the moon Enceladus, giving rise to speculation that the moon might be a source of the ring material.

Voyager 1 was already in its observatory phase, and even from millions of miles away, it had made some significant new discoveries while raising still more questions about Saturn. Beginning in January, the Voyager planetary radio astronomy experiment had been detecting regular, very-long-wavelength radio pulses from Saturn of much lower energy than the synchrotron bursts from Jupiter. These Saturnian broadcasts roughly coincided with the internal rotation of the planet's core. The regular emissions—10 hours, 39 minutes, 24 ±7 seconds—established for the first time a firm rotation rate for the interior of the planet.

The spacecraft also contributed to the swarm of new Saturnian moons, which seemed to be popping up like weeds in 1980. The newest additions, 1980S26 and 1980S27 (officially, S-13 and S-14) flanked the narrow, newly discovered F ring. S-13 orbited at a distance of 2.345 R_s, S-14 at 2.3 R_s, with the F ring tucked in between them at 2.32 R_s. The new moons were tiny, barely 200 km across, and it seemed likely that the particles in the F ring were held in their narrow orbit by the gravitational forces of the two satellites. Because of this effect, the moons soon became known as the "shepherding satellites."

Project Scientist Ed Stone welcomed the press back to JPL on November 6. Two days earlier, Ronald Reagan had been elected

---

* Named after the French mathematician Joseph Louis Lagrange, who first studied the problem, the Lagrangian points are locations in space where a third body can maintain the same position relative to a second body in orbit around a first body. The most stable of the five Lagrangian points, L-4 and L-5, are situated 60 degrees ahead of and behind the orbiting second body. In the Earth–Moon system, for example, the best place to put a permanent space station would be at L-4 or L-5. In the Saturn–Dione system, the new satellite was found at the L-4 point.

President, and with the long political season finally at an end, the world was more than ready for an extraterrestrial diversion. The press room in Von Karman Auditorium was more crowded than ever; PBS broadcast daily live reports from JPL; *Cosmos* was appearing weekly and attracting large audiences; and the Space Shuttle was at long last being readied for its first flight in the spring of 1981. In the face of the gloom and doom pronouncements in Tucson, it seemed that public interest in space was at an all-time high. The public's enthusiasm didn't solve NASA's budgetary problems, but all the publicity and excitement were heartening for the space community. Maybe a spectacular Saturn encounter coming at this propitious moment could rescue the planetary program.

Saturn did not disappoint. As Stone made clear in the initial press briefing, much less was known about Saturn than was known about Jupiter on the eve of the first Voyager encounter. Surprises were inevitable; no one realized just *how* surprising the surprises would be.

Saturn itself seemed to be behaving like a proper gas giant, although already some disturbing dissimilarities from Jupiter were becoming apparent. "Our longstanding theories . . . are likely to become casualties," said Brad Smith. "I think we can predict that much of what we thought to be true will be wrong." That may have been the most accurate prediction of the entire mission.

It was now early summer in Saturn's northern hemisphere. The atmosphere did not look markedly different than it had in Earth-based photos. "Saturn doesn't have all those nice psychedelic features," said Smith. "In fact, it looks rather bland." However, Voyager's sharp-eyed cameras were picking up some interesting new details. "Pioneer 11 scared us to death," Smith admitted, but Voyager was finding a multitude of detail in the low-contrast clouds. The familiar blobs now numbered in the hundreds. There were also some Jovian-style high-speed jets, Saturn's being unexpectedly much stronger than Jupiter's. Jupiter's equatorial jet stream moved at about 100 meters per second; Saturn's were observed zipping along at 400 mps. Such differences between the two gas giants were a little unsettling because a major mystery at Saturn could have a boomerang effect on existing Jupiter theory. As Smith put it, the Jovian theories had to apply to Saturn as well "or we're in trouble on Jupiter."

From 11 million miles away, Saturn's atmosphere showed more contrast and detail than was visible through Earth-based telescopes. But compared with Jupiter, Saturn's atmosphere seemed calm and placid.

The planet might be similar to Jupiter, but its family of moons —new and old—was markedly different from the Galilean satellites. Jupiter's moons were unique, but they varied systematically in brightness and density. At Saturn, however, there seemed to be little rhyme or reason in the ordering of the satellite system. Voyager would see an entirely new class of objects, an intermediate size range of moons between 400 and 1,500 km in diameter. Only Titan, the outsized maverick moon, was in the same size range as the Galilean satellites. The other moons were small, not very dense, and were probably composed mainly of ice, although even that information was uncertain. "We really have a very rudimentary knowledge of the Saturnian satellites," said Ed Stone.

Although the moons and the planet were fascinating, inevitably

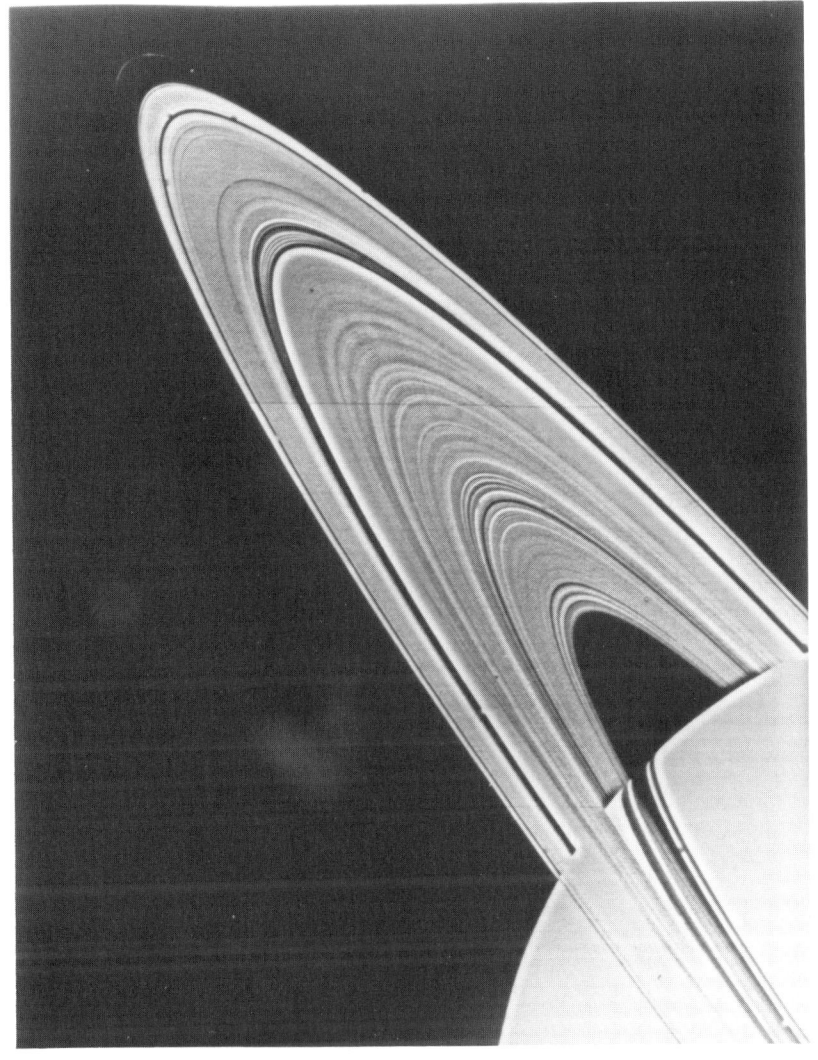

In this two-image mosaic from 4.9 million miles, some 95 concentric "ringlets" can be seen. It was soon obvious from images such as these that the classical explanation for the structure of the rings was inadequate to account for the multitude of ringlets.

Saturn's fabled rings occupied center stage. Early results from the far-encounter phase of the flyby had already effectively destroyed the existing theories about the rings. Ever since the days of Maxwell, astronomers had been plotting gravitational resonances with the satellites to explain the ring structure. But those carefully calculated resonances were based on the presumption that there were only a handful of distinct rings—A through G, now, although the D ring was still doubtful. Voyager's cameras had already found not six or seven rings, but *hundreds*.

The broad, bright C, B, and A rings turned out to be composed of many narrow "ringlets", especially in the C ring. "There's an enormous amount of structure in there," said Brad Smith. There was even obvious structure within the Cassini Division. "I think we have to admit," said Smith, "that the full question about satellite resonances forming the rings is not going to work anymore. We have to look for an alternative explanation."

The search for alternatives was going to be more difficult than anyone imagined; the ring saga would be worthy of Tolkien. On the long approach to Saturn, Voyager had taken "movies" of the rings which revealed features of surpassing strangeness—the spokes. These long, dark, fingerlike projections extended almost completely across the B ring. According to everything the scientists thought they knew about rings, the spokes were flatly impossible. Yet there they were. Dave Morrison summed up the situation: "The fact is that every year for the last decade, we have understood *less* about planetary rings than we did the year before." Torrence Johnson described the scientific reaction to the discovery of such baffling new objects as the spokes: "Observationalists say, 'Oh boy!' and the theoreticians say, 'Oh, damn!'"

Theoreticians were probably saying a lot more than that. According to Smith, the spokes were causing them "enormous problems." The problems centered on the fact that the spokes stretched radially across the B ring, like the spokes of a wagon wheel radiating from the hub, which was Saturn. The spokes had to be composed of the same stuff as the rings themselves—small particles—but they behaved in a very unringlike fashion. A particle orbiting at the inner end of the spokes would have an orbital period some three and a half hours shorter than a particle at the outer end. The shear created by the differing orbital speeds should have torn the spokes apart as soon as they formed; indeed, there was no obvious reason why the spokes should have formed in the first place. Why should some ring particles preferentially arrange themselves into spokes while the rest of the ring particles did not? Could they be made of different material? If they were, said Smith, "that gives us even more grief."

"This is quite clearly the most baffling surprise we've come up with yet in this mission," Smith told the reporters. "The answer to all of this is that we don't know. We don't even have a model. . . . I can't think of anything at Jupiter that would leave us this uncertain." When a journalist asked him why the rings and spokes were so complex, Smith stared at him in honest amazement. "You're asking me *why?*" Smith asked. *Why* was obviously far down on the list of ques-

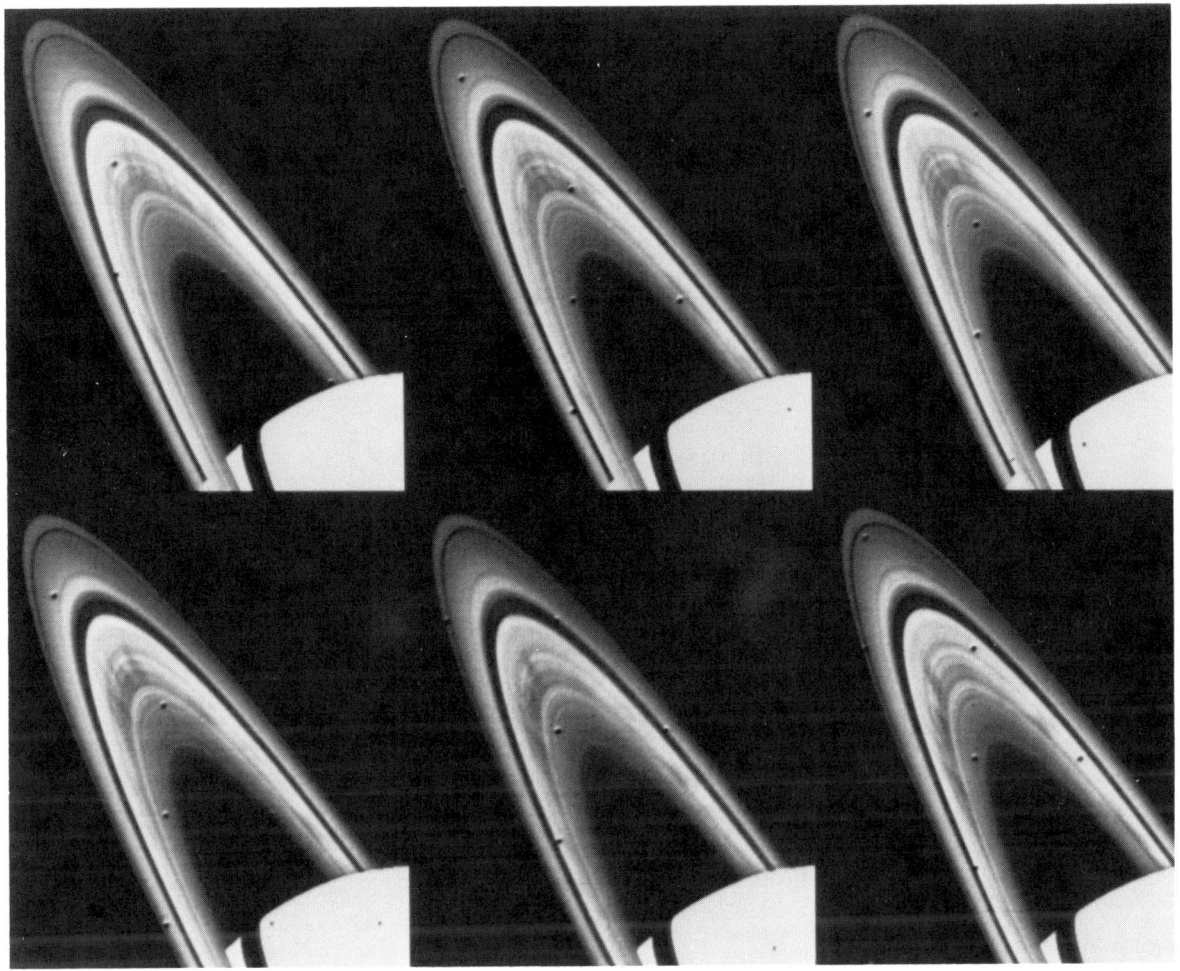

This sequence, running from upper left through lower right, shows the motion of the spokes around the rings. The images were taken every 15 minutes, from a distance of 15 million miles. Note that the structure of the spokes begins to break up with time.

tions to be asked and answered; the scientists were still only in the *what* stage.

The spokes and multitudes of ringlets served as a fitting introduction to Saturn, rather like Rod Serling's "signpost" at the border of the Twilight Zone. For scientists and laymen alike, the Voyager 1 Saturn encounter was to be a surreal journey through an unimaginable landscape populated by strange and miraculous creatures. Nothing yet seen at Jupiter—or anywhere else in the solar system or the rest of the universe—prepared earthlings for the wonders of Saturn. After one particularly mind-twisting press briefing, a reporter (and a science-fiction writer by trade) was seen walking from the auditorium, shaking his head and muttering loudly, "Toto, I have a feeling we're not in Kansas anymore!"

It may have been the best encounter of them all. Connoisseurs of such moments like to recall the day Viking 1 landed on Mars and Voyager 1's fantastic flyby of Io. But for sheer, sustained weirdness,

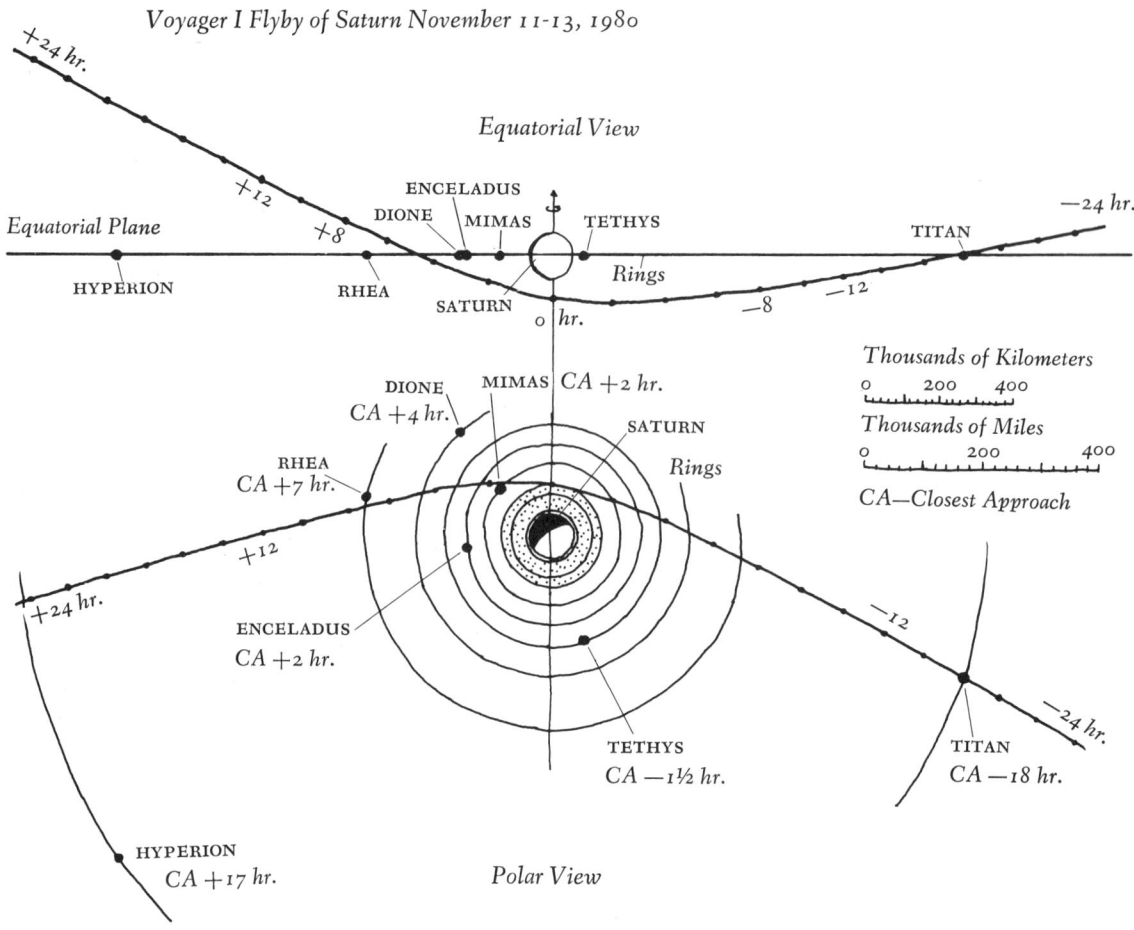

Voyager 1's trajectory at Saturn features two crossings of the ring plane and an extremely close encounter with Titan.

for unadulterated awe and mystery, there has never been anything quite like the Voyager 1 Saturn encounter.

Part of the thrill of Saturn was due to the fact that the Saturnian system is smaller than that of Jupiter. At Jupiter, the closest approaches to the satellites were spread out over several days. At Saturn, the near encounters were concentrated in a span of several hours during November 12. In the press room, and in the imaging team area, clusters of interplanetary sightseers gathered in front of the monitors, mission timelines in hand, waiting for each new set of pictures from worlds never before seen. Suddenly, someone would shout, "Here it comes!" and all eyes were riveted on the screens as the computer printed out the new image's numerical data on the right-hand side of the tube. Then the picture formed, materializing top to bottom, revealing yet another dazzling, baffling sphere. An army of forefingers attacked the monitors, jabbing at this feature or that, as experts and amateurs, veterans and rookies, turned to one another and asked, over and over, "Is this real?" As the sequence continued, theories blossomed and died with each passing minute—"instant science" had never been quite so quick or so exhilarating. Finally, the sequence completed, people checked their timelines again and sat down to wait for an hour or two, until the next checkpoint on the cosmic itinerary. Then the whole show would repeat itself—with brand-new scenery.

In the days before the closest approaches (Titan on the night of the 11th, Saturn and a flock of moons on the 12th), the scientists speculated and theorized and attempted to integrate each new nugget of data into some overall framework that made sense. There were so many distinct areas of interest about which so little was known that questions multiplied like medflies; answers were in short supply.

Saturn was looking less bland as the spacecraft got closer. Computer enhancement techniques were bringing out a wealth of detail in the golden cloud tops. The images showed more and more Jovian-style spots, including a Saturnian Great Red Spot. This spot, known as "Anne's Spot" after Anne Bunker, the imaging team worker who had tracked it carefully, was neither as great nor as red as the spot on Jupiter, but it was the best Saturn had to offer. Relative to the size of the planet, it was comparable to the large white ovals on Jupiter. It was some 10,000 to 12,000 kilometers across, and Smith said that "it would be tempting to believe" that it was anticyclonic like the Jovian spots. Other Jovian similarities were becoming apparent as a profusion of belts and zones revealed themselves, particularly in the high latitudes.

There were also some disquieting differences from Jupiter. On Jupiter, the high-speed jet streams were always situated at the boundaries of the belts and zones. On Saturn, the wind velocities at the boundaries were almost zero. The high-speed jets seemed to be at the *center* of the bands. "We just really don't understand it," said Smith. "It's more difficult to determine just what a belt or zone is . . . we may have tried to oversimplify things from Jupiter."

The location of the high-speed flows had implications for the interior structure of the planet. "It's causing us much consternation," Andy Ingersoll admitted. Ingersoll's nested-cylinder theory (the astrophysical model) and Gareth Williams' computer-generated cloud flow simulation (the meteorological model) were still the sub-

The low contrast of Saturn's clouds is shown in this image from 1 million miles. Faint bright spots were used to track atmospheric currents. The black spot at bottom is the shadow of the moon Dione.

jects of heated debate. How did the high-speed jets relate to models of upwelling and downwelling in the belts and zones? "Right now," said Ingersoll, "we have no clue."

While the imaging scientists were scratching their heads, the rest of the scientists were encountering puzzles of their own. The torus of neutral hydrogen in the vicinity of Titan's orbit turned out to be much more distended than anyone anticipated. The total mass of the torus was about what had been predicted from earlier models, but the distribution of the mass was different from expectations. One implication of the ultraviolet spectrometry data was that Titan's atmosphere might turn out to be quite dense.

Titan's atmosphere was the prime science target of the encounter. The spacecraft's trajectory had been designed to assure an extremely close pass at Titan on the inbound leg, before the hazardous ring-plane crossing. For Titan, imaging was of secondary importance, since it seemed unlikely that the moon's clouded atmosphere would permit a glimpse of the surface. The key experiment would be the occultation of the spacecraft's radio signal as it passed behind Titan. By analyzing the manner in which the signal was distorted and ultimately cut off by the moon, scientists hoped to produce a "profile" of Titan's atmosphere showing changes in temperature and pressure from the topmost clouds all the way down to the unseen surface.

The methane in the atmosphere, discovered by Kuiper in 1944, was still the source of endless speculation; no one knew how much of it there was, how it got there or what other gases might be present. The abundance and identity of organic molecules in the atmosphere might also be determined.

Two models of Titan's atmosphere were still alive, but at least one of them was likely to perish in the immediate future. In the low-pressure model, the atmosphere was almost entirely methane and the surface pressure would be no more than about 20 millibars (surface pressure on Earth is 1,000 millibars, or 1 bar). The surface temperature would be about 80°K, while upper atmosphere temperatures at the 1-millibar level might be as high as 160°K. In the high-pressure model, methane was supplemented by some other gas, probably nitrogen, giving a surface pressure as high as 2 bars and a surface temperature greater than 80°K. In terms of atmospheric chemistry, the high-pressure model was much more interesting because, as Toby Owen put it, "Chemistry at 77°K happens very slowly." "If the temperature and pressure are much higher," added Andy Ingersoll, "then we begin to entertain more exotic things . . . ammonia, water, and who knows what?"

The "who knows what" could conceivably include some biologically interesting hydrocarbons. Ground-based data indicated that there was a dense hydrocarbon smog in the Titanian atmosphere, and it seemed possible that this organic matter could be raining down onto the surface, building up a kilometer-thick layer of sludge—in which all sorts of reactions might be taking place. Further speculation depended on the precise identity of the organic smog particles.

While the atmospheric scientists speculated about what lay under Titan's blanket of smog, imaging team scientists were waiting for their first close look at the rest of Saturn's satellites. Aside from orbits and albedos, almost nothing was known about them; that didn't prevent the scientists from talking about them. "It's always easy to classify objects when you don't know much," Dave Morrison explained. The smallest of Saturn's moons were about the same size as large asteroids, and some of them—particularly Phoebe, the outermost—may have been asteroids or extinct comet nuclei captured by Saturn.

In contrast to Jupiter, where the inner satellites were rocky and the outer moons icy, most of Saturn's inner moons appeared to be icy, while the outer ones looked dark and rocky. Prevailing theory suggested that as the proto-Saturn slowly contracted, the inner moons formed late. The outer satellites collected most of the rocky material and the inner moons were left with only the volatile gases as building materials. However, the precise timing of this scenario is critical; Saturn's contraction had to have been essentially complete by the time the sun entered the T-Tauri stage, when ferocious solar winds blew most of the leftover gases out of the solar system.

As far as anyone had firm expectations, most of the scientists believed that the icy satellites would differ from the icy bodies around Jupiter. Torrence Johnson thought that, in general, Saturn's moons would turn out to be "bright, clean, icy satellites, as opposed to dirty things like Ganymede." The smaller Saturnian moons seemed unlikely to have generated much internal heat, implying a relatively inactive surface geology. Asked about Europa-style resurfacing processes as an explanation for the bright, clean surfaces, Johnson said it was more likely that the cleanness was simply due to the original composition of the satellites.

The Saturnian moons, a billion miles from the Sun, were 90° to 100°K colder than the Galilean satellites. At that temperature, ice could be as rigid as steel; but it could also be very plastic. Little was known about the behavior of ice or any other material at these extremely low temperatures. Impacts on Tethys might look very different from impacts on Callisto. "Right now," Morrison said on the 7th, "they could turn out to look like almost anything."

The small, newly discovered satellites (or newsats, as the computer displays labeled them) were equally mysterious. "We know nothing about them physically," said Morrison. The best guess was that they would turn out to be icy objects. Some rapid reprogramming was being done to get a few high-resolution images of the tiny bodies, but uncertainties about their orbits made that something of a hit-or-miss operation.

Whatever they might look like, the small co-orbital moons, S-10 and S-11, were fascinating objects from the point of view of celestial mechanics. Sharing the same orbit but traveling at different speeds, the two satellites seemed to be in danger of collision. They apparently avoided such a mishap by a neatly choreographed orbital minuet. The trailing moon was in a very slightly lower orbit (the difference in the two orbits was less than the size of the moons) which meant that it moved faster, relative to the second moon. As it caught up with the leading moon, its gravitational pull would slow down its partner. The slight deceleration would cause the leading moon to drop to a lower orbit; meanwhile, the trailing moon—accelerated by the pull of the leading moon—would shift to a slightly higher orbit. In its larger orbit, the trailing moon then began to fall behind. Eventually, the leading moon, in its lower orbit, caught up to its partner from behind and became the new trailing moon—at which point the whole dance repeated itself.

The discovery of so many new moons seemed to have jaded the scientists just a bit. "I think by the end of next week," said Torrence Johnson on the 7th, "a new satellite isn't going to be big news." Perhaps, but it was certainly news the next day when Brad Smith announced the discovery of the fifteenth Saturnian moon.

"It's in a remarkable position," said Smith. Orbiting just 800 km outside the edge of the A ring, S-15 was the newly crowned innermost Saturnian satellite. It seemed a good guess that the tiny new moon (it measured barely 100 km across its longest axis) helped to gravitationally bound the A ring. The discovery of a moon in such a position, combined with the proliferation of rings, started the scientists wondering if there might be many such moonlets hiding within the rings, creating the structure which could not be explained by the classical satellite resonances.

Such an explanation seemed necessary now because the rings were getting far too complex to be explained by any existing theory. At least ninety-five concentric rings were visible, and images of the rings were looking more and more like a very-long-playing phonograph record. Five rings showed up within the Cassini Division; "It's getting hard to tell just where the Cassini Division begins and ends anymore," commented Brad Smith.

Viewed in ultraviolet wavelengths, the structure of the rings was even more complex. In computer-processed UV images, the C ring looked bluer than the others, indicating a different particle size. The "blue" particles also appeared in the Cassini Division. "Whatever is producing the particles in the Cassini Division," said Smith, "the particles in the C ring may be similar." It was certain, at least, that in different parts of the rings the size and composition of the ring particles varied substantially.

The closer Voyager came to them, the stranger the rings looked. "The ring mystery keeps getting deeper, till we feel like it's a bottomless pit," Smith said on the 11th. The latest addition to the mystery was the discovery of two eccentric rings. All the other ringlets had concentric orbits, but these mavericks somehow moved in eccentric, elliptical paths. One of them, in the C ring, had an orbit which varied by about 60 km. It was near a possible resonance point with Mimas, but not near enough. "We may have some difficulty in using Mimas to explain this," said Smith.

Meanwhile, imaging team member Richard Terrile had found evidence of clumpiness in the F ring. In some images, the narrow ring was seen to have thick, bright knots of material scattered around its circumference. The clumps might have hidden small moonlets that were providing material to the ring, but the problem with that theory was that sooner or later the moonlets would erode to nothing and the ring would dissipate. It seemed more likely that the hypothetical moonlets, like the shepherding satellites, were simply focusing the F-ring particles rather than creating them. "It's going to be a strange animal," Dave Morrison had said of the F ring. He was right, but at the time no one had any inkling of just how strange it would turn out to be. Within a few days, the F ring's reading on the Strangeness Index was to jump right off the scale.

And then there were the spokes. "I don't think there is anything that has kept us puzzled for so long," Brad Smith had said on November 9. Usually, unexpected discoveries inspired new ideas within a few hours, but the scientists had been staring at the spokes for more than a month without coming up with any plausible explanations. "We're no closer now than we were a few weeks ago," Smith admitted. He added that he didn't expect any answers in the near future.

"We're still pretty much in the gee-whiz stage," James Pollack said of the spokes on the 11th. Rich Terrile was the imaging team's main "spokesman," and he had been trying to understand them ever since they were first seen, without notable success. "There's a high degree of randomness in these spokes," he said. They appeared suddenly as the ring particles moved out of the shadow of Saturn, and that seemed promising—perhaps sunlight had something to do with the creation of the spokes. But the spokes might also exist in the shadows where they could not be seen, and the uncertainty cast doubt on the sunlight hypothesis. For a time, it was thought that the spokes might actually *be* shadows, produced by small particles that had somehow levitated out of the ring plane. However, the sun angle was wrong for such an explanation, and the idea was discarded, along with a good many others.

One of the most perplexing of the spoke mysteries was the fact that they appeared only in the B ring. "The problem there," explained imaging scientist Allan Cook, "is to have a thing that knows it should only work in ring B." If the spoke effect was, for example, electromagnetic in nature, then presumably it would have been apparent in all of the rings and not exclusively in the B ring.

However, there was some question as to whether the spokes really were confined to the B ring. That was where Voyager saw them, but it now seemed that other observers may have seen them in the A ring years earlier. The early-twentieth-century astronomer E. M. Antoniadi had produced drawings showing spokelike features in the A ring. Although Antoniadi's reputation for accuracy was good, some Voyager scientists expressed doubts about the reliability of his "spoke" drawings.

Antoniadi might be dismissed, but the observation made by a young amateur astronomer in Cambridge, Massachusetts, were not so easily brushed off. In 1977, Stephen O'Meara, an editor at *Sky and Telescope*, made drawings of things that looked like spokes in both the A and B rings. His observations were made through a 9-inch refracting telescope at Harvard, while professional astronomers using some of the world's largest telescopes had never seen anything that resembled the spokes.

When *Sky and Telescope*'s J. Kelly Beatty arrived at JPL for the encounter, he mentioned that O'Meara may already have seen the mysterious spokes. The news traveled quickly, and an hour later Brad Smith appeared in the press room, asking Beatty, "What's this I hear about someone seeing spokes?" Smith was unconvinced by O'Meara's drawings. Lots of people had seen things from the Earth that weren't really there—Martian canals, Janus, and even the D ring, which Smith himself thought he had seen from the ground but which so far had not appeared in Pioneer or Voyager imaging or in the particles and fields data. Smith maintained that the contrast of the spokes was too low for them to have been observed from Earth, particularly with a 9-inch refractor. There was a brief flurry of publicity surrounding O'Meara's observation, but most of the scientists followed Smith's lead in dismissing them. On the other hand, O'Meara's drawings were made in 1977, when the lighting conditions on the rings were better than they were during the Voyager 1 encounter. It seems probable that O'Meara saw *something*, but whether he actually saw the spokes remains a subject of controversy.

The particles and fields men were still patiently waiting for the action to begin. Upstream solar wind data from Voyager 2 implied that Saturn's magnetosphere would be somewhat larger than it was during the Pioneer encounter. How much larger it would be was a question of considerable interest. If Saturn's magnetosphere was like Jupiter's, it could have expanded considerably; if it was more rigid, it might be only slightly larger. Either way, Titan would probably be within the magnetosphere this time.

Saturn's untilted magnetic field was still baffling the scientists, who were not sure how an untilted field could exist in the first place. "There are novel phenomena that go on because the field is not tilted," said experimenter Fred Scarf. In particular, the radio emissions Voyager had been picking up seemed at odds with the lack of a magnetic tilt. "It's very perplexing to the people who do dynamo studies," said Scarf. "We really don't understand the physics of what Pioneer saw."

The geologists, however, at least thought they understood what they were seeing as the first long-distance images of the major satellites came in. There was a bright, star-shaped pattern on Dione which seemed to suggest icy crater rays. Rhea showed patterns of dark and light which led Brad Smith to comment that "It's tempting to interpret this as a meteoric impact." Other albedo variations were seen on Tethys.

Rhea and Dione were reminding the geologists of their first views of Ganymede. Discussing the albedo pattern on Rhea, Larry Soderblom said, "We wouldn't even have hazarded a guess a year ago. It's neat to think that our terrestrial perspective has broadened to include Ganymede." Soderblom described the typical first reaction to the Dione images: "Wow! That looks like Ganymede!"

The "twisted, ropy" patterns of Rhea and Dione did, indeed, call to mind similar features on Ganymede. "We don't understand them," Soderblom explained, "but we know what they look like." The long-distance resemblance, though, was subtly misleading. Soderblom had spoken of no longer being "hampered by our terrestrial perspective," but in some ways our new Jovian perspective was also too restrictive. The superficial similarities between Rhea and Ganymede tended to obscure some fundamental differences between the two worlds. Rhea, with a diameter of 1,530 km, is less than a third the size of Ganymede. It is in a colder environment and its smaller size means that it can generate far less internal heat to drive the surface geology.

After the surprises at Jupiter, the geologists were determined not to be caught unawares by the surfaces of the Saturnian moons. Active surface processes were no longer thought of as the exclusive property of the inner planets. The star pattern on Dione, for example, immediately interested Soderblom: "It means, to me at least, there has to be some sort of internal evolution going on."

Throughout the Voyager encounters, Soderblom was quick to find evidence of "internal processing" on the moons of Jupiter and Saturn. As a geologist, Soderblom was naturally interested in anything which seemed to indicate the existence of an active extraterrestrial geology. However, the imaging team astronomers, notably Brad Smith, were more inclined to see evidence of *their* special field of interest, such as cratering. The "internal vs. external" dichotomy sharpened the scientific debate, and proved the wisdom of including both astronomers and geologists on the imaging team. The natural predispositions of each group tended to balance one another and led to a more thorough consideration of all the evidence.

Other moons were also showing some early promise. On Tethys, there was "a very, very strange-looking feature," which might have been a mound or a hill. "If it's a hill," said Soderblom, "it's a heckuva hill." Large vertical relief on Tethys would be a bit of a surprise because of the moon's icy composition, which ought not have been able to support large mountains.

Voyager 1 would not get a very close look at Iapetus, the bizarre bright–dark outer moon. The best images from Voyager 1 would come following the close encounter with Saturn, but truly high-resolution pictures would have to wait until Voyager 2 arrived the following August. The first long-range images of the moon revealed that the dividing line between the bright and dark terrain on Iapetus was surprisingly sharp. In fact, the demarcation was entirely too sharp for the theory that Iapetus was simply sweeping up Phoebe dust. "I doubt that it can be due to the collecting of junk in Iapetus' orbit as it moves through space," said Soderblom. "A better speculation is that some sort of internal process has gone on." Soderblom called the discovery "a striking surprise, and a happy one."

On the evening of November 11, Voyager 1 made its closest approach to Titan, at a distance of less than 5,000 km. It was the closest flyby of another world ever attempted, and the spacecraft hit the bulls-eye. Mission designers were hoping for an accuracy of plus or minus 265 km; what they got was an astonishingly accurate encounter trajectory that missed the aiming point by barely 20 km. Charlie Kohlhase, who used to hunt squirrels, proved once again that he and his colleagues were the Space Age equivalent of Davy Crockett, the legendary long-range sharpshooter.

During the approach to Titan, the question on everyone's mind was whether there would be any breaks in the clouds or even any detail in the clouds themselves. The early images of the moon made it look like a featureless fuzzball. Since nothing else seen by Voyager had been visually uninteresting, it seemed incredible that Titan would turn out boring. Many people simply refused to accept the obvious and insisted that within a day the images of Titan would reveal at least some detail.

Surprise was the rule rather than the exception in the Voyager mission, and probably the most surprising outcome one could imagine would be that Titan would turn out to be a bland cueball of a planet. But the images on the monitors that night were all but featureless. There was evidence of a slight thickening of the atmosphere

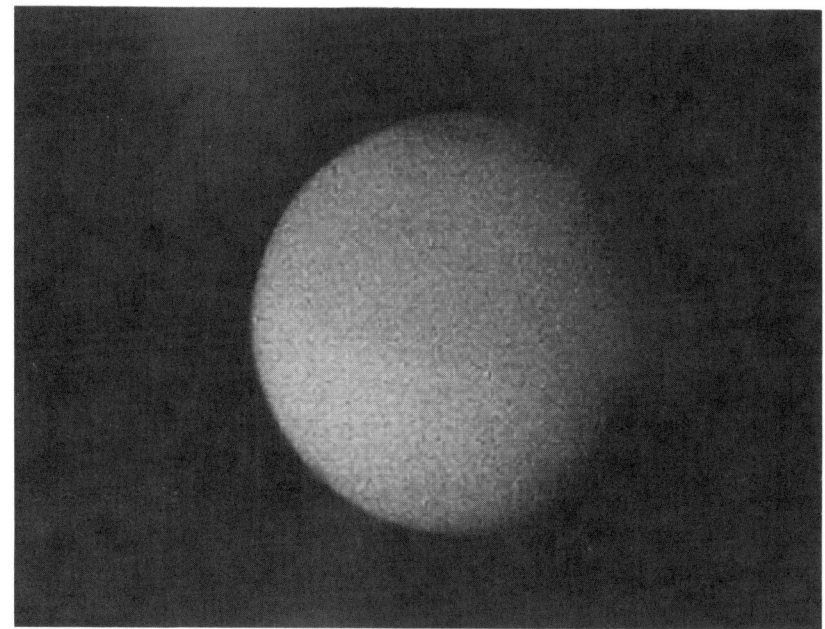

Visually uninspiring, Titan was best studied by instruments other than Voyager's cameras. The southern hemisphere is somewhat brighter than the northern. A slight darkening of clouds can be seen in the north polar region.

over the north pole, producing a "polar hood." After the images had been computer-processed, a sharp dividing line along the equator was also apparent. Titan's northern hemisphere was slightly darker than its southern hemisphere, for reasons yet unknown. Two months of additional computer processing eventually brought out some faint horizontal bands in the atmosphere, but no amount of computer stretching could alter the truth: Titan is the most visually bland world yet seen in the solar system.

It didn't really matter. The most important Titan observations were made by the radio occultation experiment. Voyager closed to minimum distance at 11:06 that evening; six minutes later, it passed behind Titan, cutting off the spacecraft's radio signal until it re-emerged from occultation at 11:25 PM. Careful analysis of the manner in which the signal was extinguished would yield vital information on Titan's atmosphere, as well as give the scientists their first accurate measurement of Titan's diameter.

Ray Heacock's first announcement at the morning press briefing on the 12th was that the Titan occultation data had come down in good shape, despite yet another rainstorm in Madrid. This one had arrived over the tracking station just a few minutes prior to the start of the occultation and stayed there until 30 minutes after it ended. But Ed Stone reported that the data was coming out clear. It would take a few days for the radio science team to analyze their results, but the preliminary data was enormously exciting.

The low-pressure model of Titan's atmosphere had died during

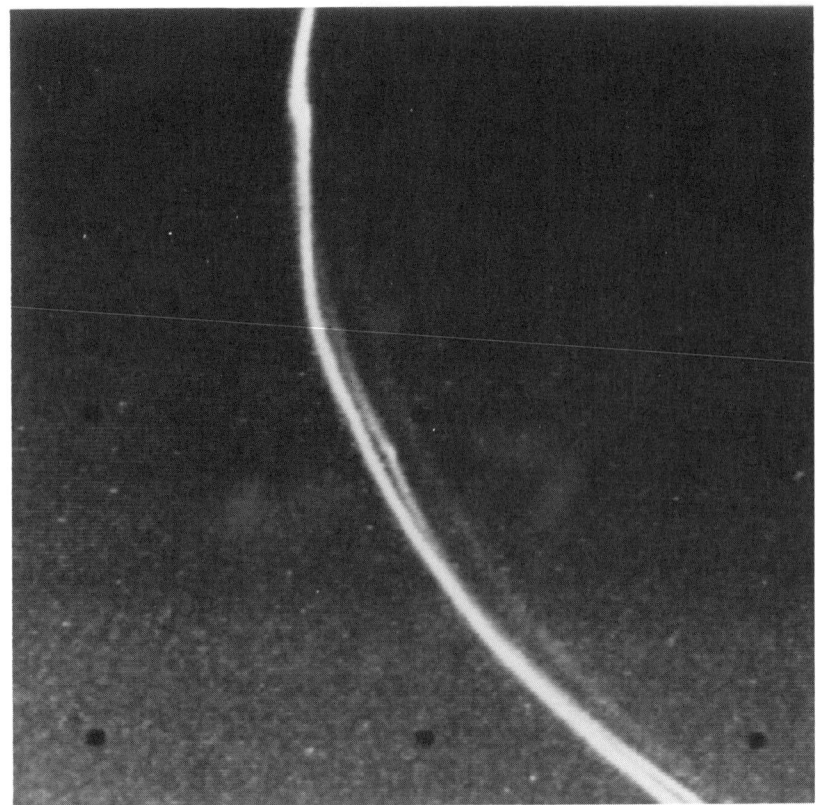

The astonishing F ring. Two bright strands and one faint, diffuse band are visible in this historic image, taken from a range of 470,000 miles. Scientists wondered if "moonlets" might be hiding in the clumpy regions, accounting for the bizarre structure.

the night. Titan's atmosphere was certainly denser than the 20-millibar model predicted. Exactly how much denser remained to be determined. The scientists were still sorting through their data and had not yet found the exact position of Titan's surface. The atmospheric density, for the moment at least, was virtually open-ended. The surface could be warmer than 80°K, possibly *much* warmer. For scientists such as Carl Sagan, who had been saying for years that Titan would turn out to be "biologically interesting," this was the moment of vindication. With much work remaining to be done, it was already obvious that Titan, bland though it may have appeared, was one of the most scientifically exciting objects in the solar system.

Shortly after the briefing, someone asked Carl Sagan for his reaction to the Titan data "in twenty-five words or less." Sagan needed just one word: "Fan*tas*tic!" as he beamed, grinning ear to ear.

It was a day for the fantastic. At the briefing, Brad Smith showed an image of the F ring that may be the most astonishing picture ever returned from a spacecraft.

At high resolution, the knots and clumps in the F ring resolved into three distinct strands of ring material. "They *appear*, anyway, to be braided," said Smith, who didn't even try to hide the disbelief in his voice. "We won't talk about this slide in the Q and A," he added facetiously.

Later, Smith wrote about the moment of discovery: "Those of us watching the monitors were stunned. If, by our third planetary encounter, we had become somewhat jaded to the unpredictability of the outer solar system, our sense of astonishment was brought back in an instant. In some tabulation of ring phenomena that we least expected to see, the observed structure of the F ring would have been somewhere off the top. In the pictures were three individual strands, each approximately 20 km wide and separated by a few tens of km; they appeared to be knotted, kinked, and braided. To me, it was the most improbable picture yet sent back by either Voyager spacecraft. . . . Until Voyager 2 can give us some idea of the time variability and stability of the F ring features . . . all of our explanations are likely to be little more than an exercise in vigorous arm waving."*

The arm waving began immediately. At first sight, the braids appeared to be woven in a tight but possibly random pattern. Could the braids be a temporary phenomenon, seen in the act of breaking up? "That's sort of untenable," said Smith. Pioneer 11, a year earlier, had seen a slight hint of the clumps. "That weird configuration is *stable*," said Smith, shaking his head.

Spokes . . . ringlets . . . moonlets . . . and now, braids! "It boggles the mind that it even exists," said Smith. "It defies the laws of pure orbital mechanics, as I understand them."

That last comment caused some minor controversy. Some journalists took it to mean that Smith thought that un- or supernatural forces were at work in the F ring. He clarified the remark a few days later: "Of *course* they're obeying the laws of physics," he insisted. The problem was that no one understood all the implications of those laws well enough to explain the braids. The scientists could only hope that Voyager 2 would somehow unravel those exasperating braids. As it turned out, Voyager 2 did just that . . . in a way no one anticipated.

Brad Smith's F ring picture was a tough act to follow, but Larry Soderblom came close to matching it. The satellite pictures were

---

* From *The New Solar System*, edited by J. Kelly Beatty, Brian O'Leary, and Andrew Chaikin. Sky Publishing Corp., Cambridge, 1981.

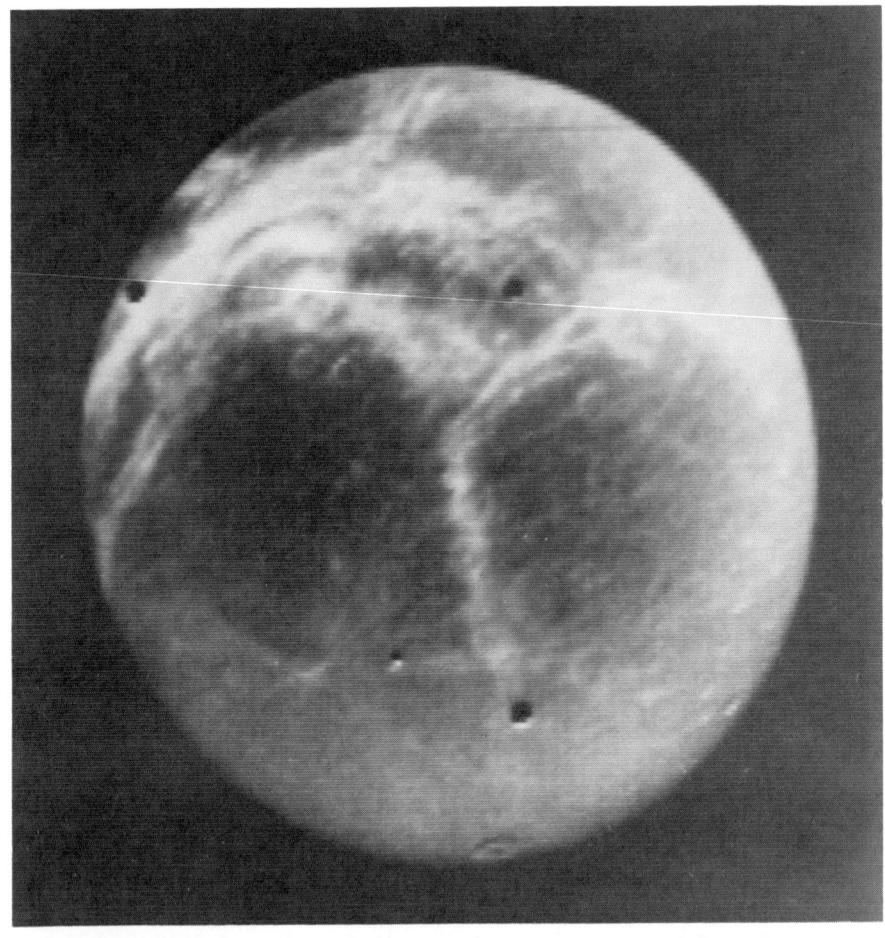

Dione (top), seen from 500,000 miles, revealed bright, wispy markings —probably surface frost. Dione is just 700 miles in diameter. Tethys (left), 600 miles in diameter, displayed a huge valley, or trench, 500 miles long and 40 miles wide. Tiny Mimas, less than 250 miles in diameter, was dominated by the huge "Death Star" crater, 80 miles in diameter. The central peak and raised crater rim are clearly visible. The ancient impact must have come close to shattering Mimas.

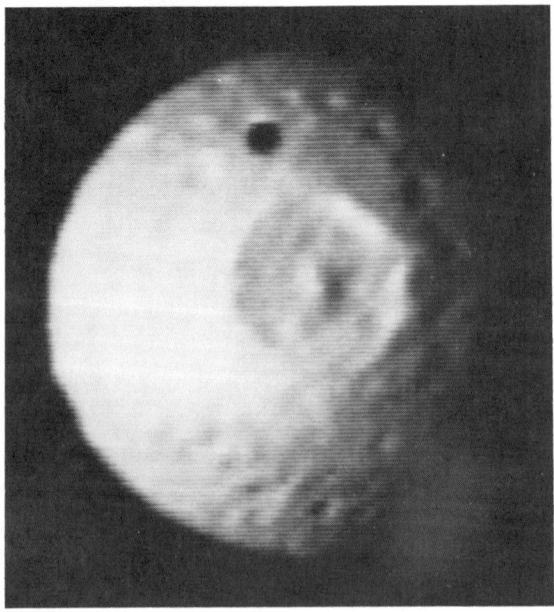

getting better by the hour. The star-shaped pattern on Dione now seemed to consist of wispy material that was connected all over the planet. On Tethys, the possible "heckuva hill" turned out to be just the opposite—an enormous trench, a kilometer deep and 50 to 100 km across. And Mimas. . .

Once more, the Von Karman Auditorium movie screen showed an absolutely astonishing image from the cosmic funhouse that was the Saturnian system. Dominating the sphere of Mimas was a gargantuan impact crater, nearly a quarter of the diameter of the entire moon. "Relative to the size of the body," said Soderblom, "this is the largest impact in the solar system." Soderblom's continuing concern about internal geological processes went out the window, as far as Mimas was concerned. "Heating Mimas isn't nearly so much of a problem as the potential for smashing it to smithereens."

The Mimas impact scar was so immense that the collision must have come very close to shattering the 390-km moon. That Mimas was still in one piece was almost shocking. Another bizarre aspect was that the moon was so hauntingly familiar, even though nothing remotely like it had ever been seen in the solar system. In fact, a body that mightily resembled Mimas had been seen "long ago, in a galaxy far away." Mimas looked like a near-clone of Darth Vader's menacing Death Star in the film *Star Wars*. The resemblance, as Brad Smith put it, was "uncanny." George Lucas (who, at the time, was busily recruiting JPL computer graphics experts for his Lucasfilm empire in northern California) must have been proud.

Almost overlooked in the excitement of the morning was the fact that Voyager had entered the magnetosphere during the late afternoon of November 11. The bowshock first appeared to 26.2 R_s. There was a total of five crossings of the magnetopause, the final one coming at 22.9 R_s—compared to 17.3 R_s for Pioneer 11. Experimenter Norm Ness called it "a classic, textbook example of a bowshock crossing." Bowshock crossings were already mundane textbook material, it seemed, even though this was only the sixth time in history that such an event had occurred.

Throughout the afternoon of the 12th, the close-encounter images from the major satellites arrived and instant science flourished once again. Tethys revealed "a crater-drenched surface," as Larry Soderblom described it. But the ancient crater field was representative of only part of the planet: "Half of Tethys has evolved geologically beyond a cold snowball," said Soderblom. "The ice at one time had to be mobile." Dione also seemed to have experienced several episodes of freezing, melting, and refreezing. Enceladus was seen only from a great distance; at maximum resolution, not a single impact crater was visible. Enceladus was increasingly reminiscent of Europa, with the list of similarities including such items as the fact that Voyager 1 didn't see either moon up close, both are the second major satellite of their planet, and both begin with the letter "E."

Late in the afternoon, the narrow-angle camera provided a detailed image of one of the co-orbitals, S-11. Previously seen end-on, it now looked like an elongated chunk of some larger body. "It's tempting to suspect that they [the co-orbitals] may have been a single body," said Brad Smith. If Mimas narrowly escaped destruction, perhaps the co-orbitals were remnants of a moon that wasn't so fortunate.

Meanwhile, Titan rumors were proliferating like ringlets. The radio occultation data were still being analyzed, and, according to the rumor mill, the surface still had not been found and the atmospheric pressure was continuing to rise. One rumor had it that the temperature in the low atmosphere was as high as 200°K. The implications of such a high temperature were staggering—all sorts of interesting chemistry could occur at that relatively warm reading. However, the rumors were premature, and it would turn out that Titan was not quite so balmy after all.

Beginning on the 13th, the experimenters presented their preliminary science results to the press. All the usual "instant science" caveats applied, and a few of the early post-encounter theories later turned out to be quite wrong. On balance, however, the scientists did a magnificent job of interpreting their multibillion-bit bonanza from Saturn.

The early Titan temperature and pressure results were presented by IRIS PI Rudy Hanel and radio science investigator Von Eshelman. Hanel reported that methane "can only be a minor constituent in the atmosphere." The bulk of the Titanian atmosphere is composed of nitrogen. The presence of so much nitrogen means that the chemical reactions occurring in the upper atmosphere ought to have produced large amounts of hydrogen cyanide, HCN. "This is a very significant molecule," said Hanel. "It has been considered a building block of more complex molecules." Among those more complex molecules are the amino acids. The biological implications were fascinating, but Hanel cautioned that "One should not expect any living creatures there." Eshelman reported that his team still had not found the surface of Titan in their data, but the atmospheric pressure had been measured to a level of 380 millibars. The best estimate was that the surface pressure "is on the order of 1 bar or more, where the 'more' is open-ended at the present time."

These initial results were tantalizing. Atmospheric expert Andy Ingersoll said of Titan, "It's a fantastic place." Conditions on the surface could be wonderfully exotic. "There could be puddles, pools, or oceans of liquid nitrogen," said Ingersoll. "It's just too bad we can't see them." But Ingersoll was mainly interested in the possibilities of the nitrogen-rich atmosphere: "The amazing thing to me is that you take four simple building blocks—hydrogen, carbon, nitrogen, and oxygen—and you can form any planetary atmosphere from them ... and yet, these atmospheres differ."

Ultraviolet experimenter Darrel Strobel described the details of some of the possible interactions among those building blocks. Energetic electrons from the magnetosphere could be colliding with upper-atmosphere nitrogen molecules, creating a population of free nitrogen ions, which could react with the methane to produce HCN. The hydrogen cyanide would then react with sunlight photons to produce a variety of long, complex organic molecules called polymers. The Titan haze layer was probably composed of these HCN products, which were reddish brown in color. The structures of these polymers could be extremely complex, as could the chemical reactions in which they participated.

Norm Ness reported that there was no evidence for a Titan magnetic field. The hypothetical wonders of a field within a field could be discarded. However, Titan did create its own magnetic wake as it moved through Saturn's magnetosphere. The interaction of Titan and the field created an induced electric current with a total output of some 20 megawatts. Ness said this was "minuscule" in comparison with the electrical energy created by Io, but it was still an example of "a power station naturally occurring in our solar system."

Although it had no magnetic field, Titan, according to planetary radio astronomy investigator Dale Gurnett, "appears to be a moderately powerful radio source." But this was "a totally new and different type" of radio emission. The source region seemed to be well above the surface of the moon, on the Saturn-facing side. The best early guess was that the radio emissions were produced by oscillations in the plasma surrounding the moon. Similar emissions, at a very low level, had been detected coming from the Earth itself; but their discovery at Titan was unexpected. The existence of such a radio source had some very broad implications for all radio astronomy. Here was one more kind of signal that could be generated in the immense physical laboratory of the cosmos.

At the time of the final briefing, on November 17, the scientists were still plumbing the depths of Titan's atmosphere, looking for the surface. They hadn't found it precisely, though they had managed to put an upper limit on Titan's radius. The moon could be no more than about 2,560 km in radius, which makes it slightly smaller than Jupiter's moon Ganymede with its radius of 2,635 km. Titan was unceremoniously dethroned as the largest moon in the solar system. Interestingly, Titan's density seems to be almost exactly the same as Ganymede's—about 1.9. Both moons are apparently composed of a relatively equal mix of rock and ice.

The early results suggested that the atmosphere was almost entirely nitrogen, with less than 1 percent methane. But further analysis showed that the true nitrogen abundance is more like 85 percent. Methane accounts for as much as 6 percent near the surface, but less than 2.6 at higher altitudes. Argon, an inert gas, seems to make up about 12 percent of the atmosphere. Significantly, neon, another inert gas, seems to be almost completely absent. If Titan had simply captured its atmosphere from the remnant solar nebula, neon ought to have been present. Its absence implies that Titan's atmosphere was not captured; rather, it had been outgassed from interior rocks and ices.

Eventually, the surface pressure was determined to be 1.6 bars, making Titan's atmosphere considerably denser than Earth's. The surface temperature was set at 93°K, although warmer regions were detected in the upper atmosphere. The combination of all these factors made Titan an incredibly interesting body. Reconstructing the evolution of Titan's atmosphere led to a scenario in which much of the original atmosphere was composed of ammonia. With an abundance of ammonia, Titan would have had a modest greenhouse effect, as the ammonia trapped heat from the sun and the warm proto-Saturn. "Some very interesting chemistry may have taken place at that time," said investigator Toby Owen.

Today, the temperature is lower, so the chemistry may be a bit less interesting. But the 93°K surface temperature is high enough to permit a variety of reactions within the atmosphere and in the layer of organic aerosols that have been raining down onto the surface. The 93°K reading is also significant because it is near the triple-point of methane, the temperature at which it can exist in three different states—gas, liquid or solid. In such an environment, methane could play the same role that water plays on Earth. There could be dense methane clouds floating above methane seas filled with methane icebergs (which would sink rather than float), with methane waves lapping against a beach of thick organic sludge. "Titan is an extremely alien surface," said Toby Owen.

Voyager 1 was hardly past Titan before the scientists began dreaming of a Titan orbiting imaging radar mission. Such a mission would be the only means of getting a look at that alien surface, outside of an even more expensive Titan lander. Given the current climate in Washington, no one had much hope of getting either an orbiter or a lander or, indeed, any sort of Titan mission within the foreseeable future.

Saturn itself had turned out to be maddeningly different from Jupiter, for reasons not immediately obvious. The Saturnian zonal jets average about four times faster than those on Jupiter. The peak zonal velocity is about 1,000 miles per hour at the equator. This eastward current falls off smoothly until 40° latitude, where the speed of the current relative to the internal rotation is effectively zero. At higher latitudes, there were more belts and zones apparent than on Jupiter, despite the relatively blander appearance of the planet as a whole.

Citing the proliferation of ringlets, gaplets, and hypothetical moonlets, Andy Ingersoll suggested that Saturn's atmosphere might more properly be said to consist of "beltlets and zonelets." However, even the classically identified belts and zones seemed to have little to do with the jet streams, which maintained their position in spite of changes in the appearance of the cloud tops all around them. "You

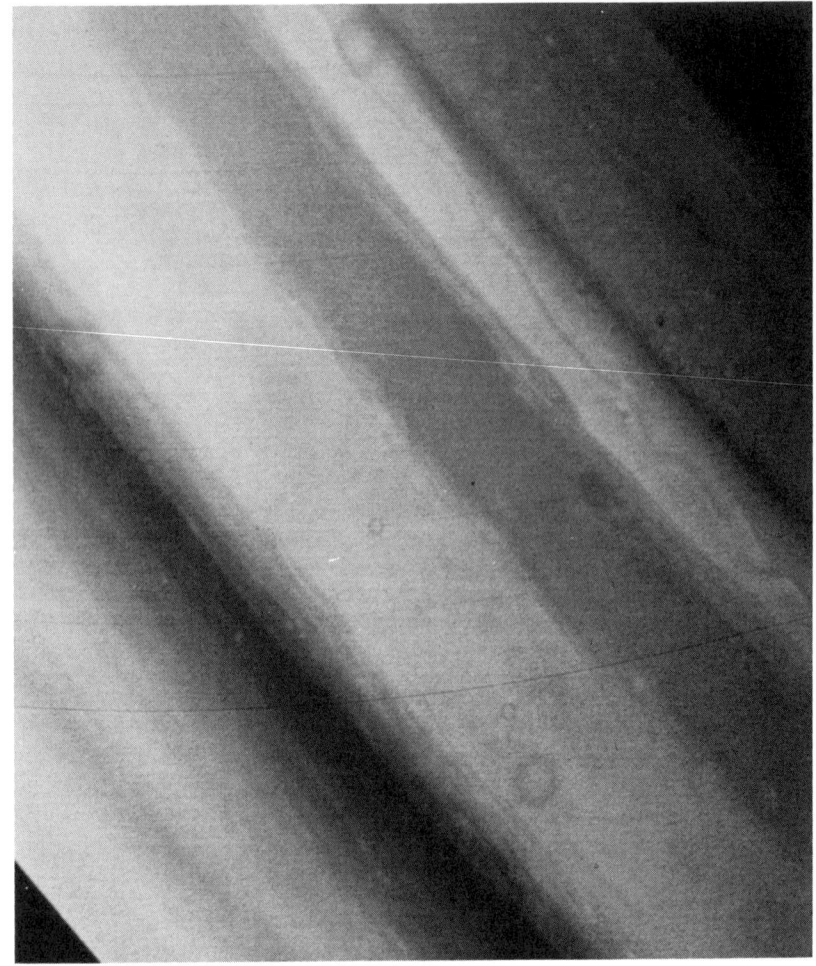

This enhanced image of Saturn's northern hemisphere shows oval structures reminiscent of oval clouds in Jupiter's atmosphere. These ovals are about 6,000 miles across. A dark "jet stream" meanders at upper right. Unlike Jupiter, on Saturn the jet streams are in the center of the belts and zones rather than at the boundaries.

have been victimized by human beings' tendency to classify things," investigator Reta Beebe told the press. "I would like to throw the belts and zones away."

The main difference between Jupiter and Saturn is that Saturn is much colder. It is also smaller (with less gravity), releases just half as much energy as Jupiter, and it has distinct seasons. All these differences combine to make Saturn a unique world, but no one could say exactly how it all fit together. "Only time will tell," said Ingersoll.

Saturn's magnetosphere seemed to be grossly similar to Jupiter's, but with specific differences created by the presence of Titan—a moon with an atmosphere—instead of Io—an airless moon with active volcanoes. At Jupiter, the complex interactions with Io tended to fill up the magnetosphere with sulfur and oxygen ions. At Saturn, the huge hydrogen torus seemed to be supplied with material from Titan's upper atmosphere. Although some extremely energetic ions had been found, moving at velocities as high as 6,000 miles per second, the overall energy level of Saturn's magnetosphere was about ten times less than for Jupiter's magnetosphere.

The non-Titan satellites were not as easily analyzed as the Galilean satellites, which seem in retrospect to be a very orderly collection of moons. The Saturnian satellites are smaller and display a wide variety of surface conditions. Most of them seem to have undergone at least some internal processing, but again, there is no obvious pattern. "We've been looking for a trend," said Larry Soderblom, but what they found was "random behavior." Soderblom suggested that we may be seeing a size range that emphasizes the random nature of planetary accretion. On larger bodies, the random processes tend to average out; on smaller bodies, the occasional random event such as the Mimas impact is more obvious. Based on crater sizes and numbers, there was also reason to think that the moon-forming process at Saturn may have been a two-stage operation. The rate of cratering and the size of the impacting objects seemed to be different in the outer solar system. Theories about accretion and cratering were based on evidence from the inner solar system, particularly Earth's moon and Mercury. To Soderblom, the differences at Saturn implied that "we started out with too simple a model."

Enceladus remains "the mystery of the Saturnian system." Part of the mystery is due to the fact that Voyager 1 didn't see Enceladus very well, but part of the mystery seems to be intrinsic. At long range, no large impact craters were in evidence. The surface looked bright and shiny and was reminiscent of similar long-range views of Europa. The Europa analogy was looking better and better. Rich Terrile and Allan Cook had done some calculations which indicated that Enceladus might be getting tidal energy from Dione and Saturn. The total amount of tidal energy could be only about one-fortieth as great as for Io, but it might be enough to power a resurfacing process.

Iapetus would also have to wait for Voyager 2 to provide high-resolution images. But the long-range shots from Voyager 1 were good enough to make the moon look even more bizarre than had been expected. The geometry of the encounter caught the bright–dark hemispheres in an orientation that resembled a huge, cosmic yin-yang sign. Right on the border between the two terrains was a very large "ringed structure." It resembled an impact, but, reported Larry Soderblom, "There's still an argument as to what's going on." The

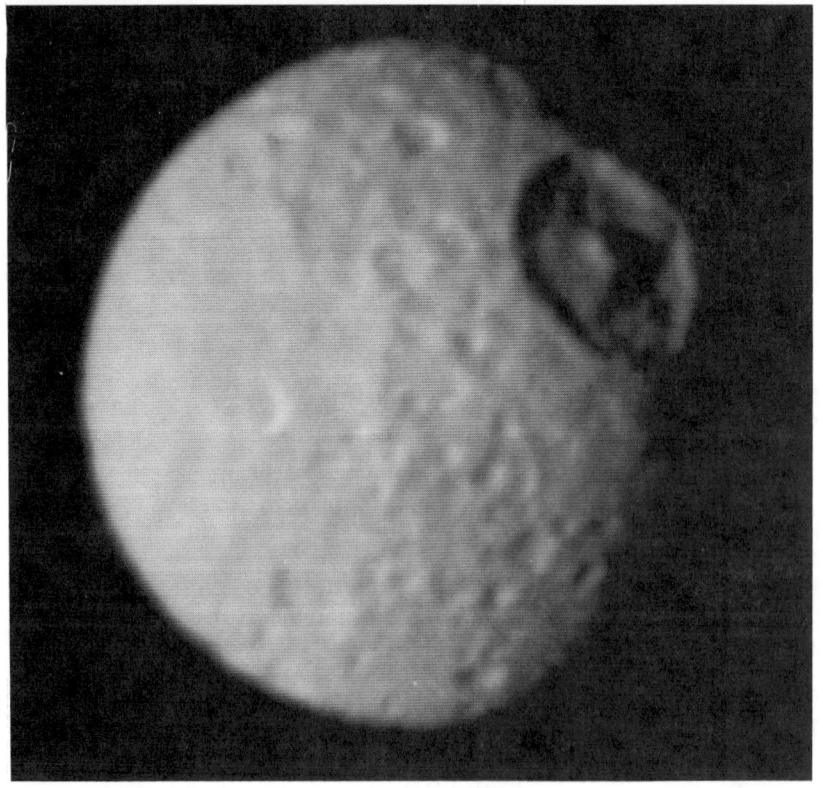

Impact craters dominate Dione (upper left) and Rhea (above). The largest crater on Dione is about 60 miles across; on larger Rhea (1,500 miles in diameter), craters up to 185 miles across are visible. The crater-saturated surfaces are similar to those of Mercury and Earth's moon. The size of the "Death Star" crater on Mimas (left) is emphasized by sunlight striking the crater wall on the night side.

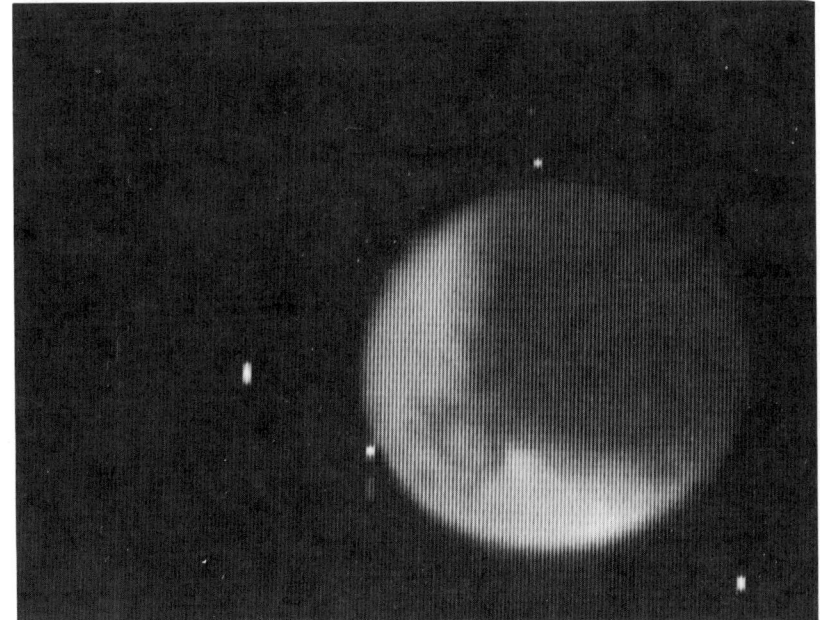

The large circular feature at the boundary between bright and dark terrains on Iapetus was intriguing: was it evidence of internal activity or a well-placed impact? Note the unusually sharp boundary between bright and dark terrain.

boundary was too sharp and the ringed feature too precisely placed for the theory that the dark side of Iapetus was simply sweeping up dust sputtered from Phoebe. The early suspicion was that Iapetus was an evolved body, and that either the dark stuff or the bright stuff was somehow coming from the interior.

The tiny newsats were still being tracked down and identified in the images. Virtually everyone expected more of them to be found. In the meantime, a pair of images of the trailing co-orbital moon resulted in the unexpected discovery of a more or less new ring. The images, taken nine minutes apart, showed a thin, dark line had moved across the surface of the tooth-shaped satellite. The line was apparently the shadow cast by a ring a few thousand kilometers from the F ring. This new ring was in roughly the same position as the G ring discovered in the Pioneer 11 particles and fields data. However, the match was not exact, so the discovery raised questions about just what Pioneer 11 *did* see. Brad Smith was not convinced that the ring shadow was produced by Pioneer's unseen G ring. As it turned out, Smith was right; the shadow was apparently cast by the F ring.

Smith was in a unique position to talk about the reality of rings which may or may not have been seen from Earth or Pioneer 11. Years earlier, Smith had seen the supposed innermost D ring through a telescope. Pioneer 11 didn't see the D ring at all. Now, Voyager had unquestionably found it—extremely faint, but definitely there. The only problem was that the D ring was *too* faint. "It's unlikely that this is an object that has been seen from the Earth," said Smith. In his

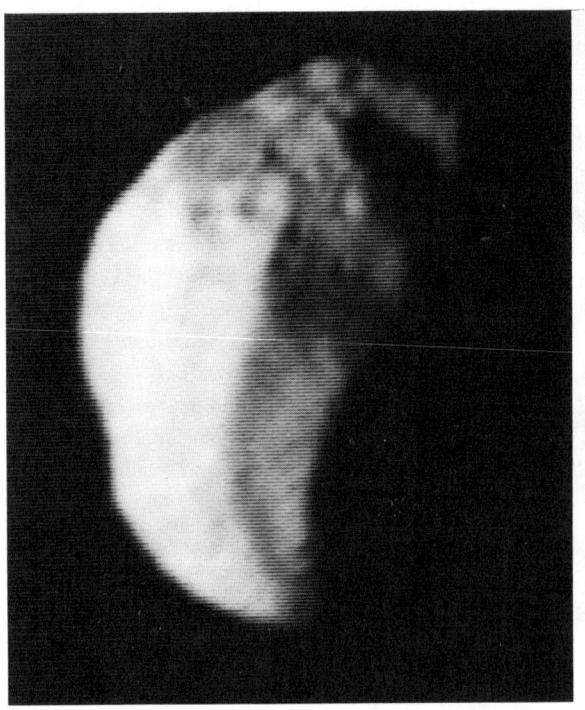 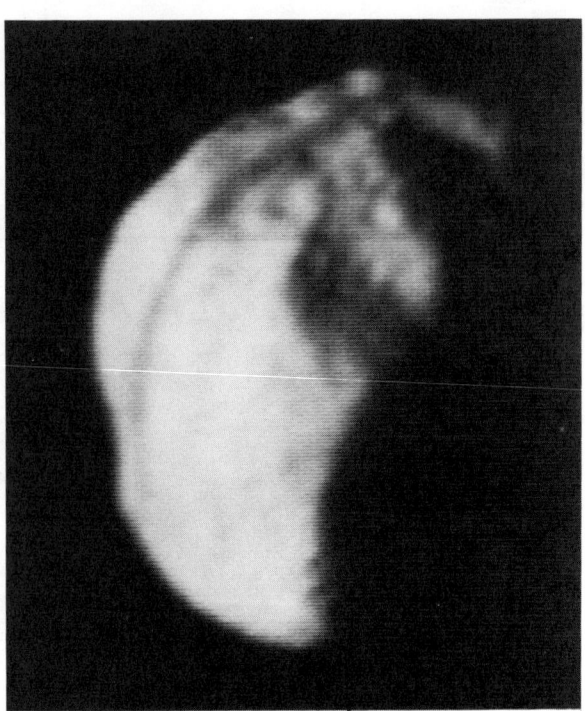

earlier observations, Smith may have been fooled by light scattered from the other rings. "It's purely fortuitous that there really is something there," said Smith.

The discovery of two new rings simply complicated the already Herculean task of making sense of Saturn's ring system. Now that the spacecraft was beyond Saturn, looking back at the rings, differences in particle size were becoming apparent in the various rings. On the sun side of Saturn, the brightest parts of the rings were composed of large particles, which effectively backscatter light. On the far side of Saturn, the large particles blocked sunlight and were dark. In-bound, the small, forward-scattering particles looked dark; out-bound, they were bright.

The A ring looked flat and uniform, much as it had always looked from the Earth. The B ring was completely different: dark, opaque, and full of structure. Ring investigator Jeff Cuzzi described it as consisting of "small stuff scattered around without any apparent rhyme or reason." The C ring had a semi-regular structure similar to the structure found in the Cassini Division. Radio occultation data revealed that the effective particle size in the C ring was about 2 meters. "In human terms," said experimenter Len Tyler, "these are not particles, these are boulders, flying around Saturn in some kind of highly organized manner."

In contrast to the inner rings, the F and E rings were composed

S-11, the trailing co-orbital moon, measures about 83 by 43 miles. It may be a remnant of a larger moon destroyed by impacts. The moving shadow line in these images, taken 13 minutes apart, may have been cast by the F ring.

The rings, seen from the night side, 8 hours after closest approach. From right: F ring, "Pioneer Gap," A ring, dark Cassini Division, broad B ring, and dim gray C ring. The spokes in the B ring, which appear bright from the night side, imply the presence of very small, forward-scattering particles.

of extremely fine material, averaging about $2/10,000$ inch in diameter. The small size of the F-ring material immediately suggested a possible solution—or at least, a path to the solution—for the problem of the mind-boggling braids. The tiny particles could be subject to non-gravitational forces. If the particles were electrostatically charged, for example, the braiding might be caused by an interaction with Saturn's magnetic field . . . or not. At this stage, no one was leaping forward with iron-clad pronouncements about the nature of the F ring.

Other aspects of the structure of the ring system were hard to explain. The classical satellite resonance theory just wasn't working very well anymore. The Cassini Division ought to have been completely free of ring particles; instead, it was filled with at least a dozen ringlets in a very regular structure. In a very broad way, the classical resonances did exist, but much else was happening.

By the end of the encounter, the number of ringlets stood at about 1,000. There simply weren't enough moons to produce such a complex structure. Speculation began to center on "embedded moonlets" —bodies 10 to 50 km in diameter that could be hiding within the rings themselves, controlling their structure and perhaps feeding material to them. For lack of any other obvious answer, the embedded moonlets became a kind of article of faith. Voyager 2 would be programmed to make a comprehensive search for them, and the scientists were confident that within a year they would have the moonlets they needed to make sense of the rings.

Some other ring phenomena were going to be difficult to understand with or without embedded moonlets. The radio astronomy experiment had detected what experimenter James Warwick described as "extraordinarily intense radio emissions" coming from the rings. The emission, at 9 MHz, was so intense that the scientists first assumed that the reading was a glitch. But the emissions, which came at a number of frequencies, were real. In some respects, they resembled terrestrial lightning flashes, but since they were originating in the rings rather than the atmosphere, analogies were suspect. It would clearly take some time to understand these new emissions.

The spokes observed in the B ring remained the major mystery of the encounter. In-bound, they were dark; now, looking back, the spokes were bright, indicating that, like the F ring, they were composed of extremely small forward-scattering particles. Again, nongravitational forces may have been at work. Rich Terrile said that the spokesmen were "slowly converging on a set of ideas" about the spokes, although they were far from a final answer. The spokes apparently co-rotated with Saturn's magnetic field. Somehow, the tiny particles may have been getting electrostatically charged, a process which could levitate them slightly out of the ring plane. But this was all very iffy, and no one was ready to claim that it was the answer. "I'm not proposing that the spokes are electrostatic interactions," Terrile said cautiously, "but they may, indeed, turn out to be."

If there was a single grand lesson to be learned from the encounter, it was that real worlds are more complex and more random than the best theoretical models. The gas giants were not well understood; the origin and nature of planetary rings had suddenly become a fantastically complex problem; and the chaotic Saturnian moons seemed to be undermining basic theory about the formation of the solar system ("Is there a cosmologist in the house?" Larry Soderblom had asked at one briefing.) Some of the early post-encounter theories would turn out to be dead wrong, as had many of the pre-encounter theories. But the data gathered by Voyager 1 would undoubtedly lead in the long run to a better understanding of the universe we inhabit.

For the moment, the scientists (and the press) were still dazed by what they had just witnessed—ringlets, gaplets, braids, Death Stars, spokes, trenches, radio emissions, jets, co-orbitals, and the atmosphere of Titan. Even Brad Smith, who is invariably referred to in the press as "laconic", allowed himself to be carried away by the marvels of Saturn. "I'm just stunned by the spectacular display of pictures," he said at one briefing. "I cannot recall being in such a state of euphoria for any previous planetary encounter."

Euphoria. The gloom and doom of Tucson were momentarily forgotten in the golden afterglow of Saturn. Writers groped for the appropriate words to describe the experience; perhaps predictably, it was Carl Sagan who found the right way to put it.

Sagan quoted a letter written in the late seventeenth century by astronomer Christiaan Huygens, who had been among the very first to view Jupiter and Saturn telescopically. Galileo had been there a half century earlier, and even if he didn't understand them, he was the first human being to gaze upon the moons of Jupiter and the rings of Saturn. Galileo, wrote Huygens, must have viewed these wonders "with no small rapture."

We can only guess, of course, but I suspect that the emotions felt by Galileo and described so eloquently by Huygens were precisely the same emotions felt by all of us at the Jet Propulsion Laboratory during those two unbelievable weeks in November, 1980, and shared by millions more on this small but special planet. Astonishment... disbelief... euphoria... and no small rapture.

10 THE LONG GOODBYE

James Montgomery Beggs looked uncomfortable. The newly appointed NASA Administrator had come to JPL to celebrate Voyager 2's successful encounter with Saturn; now he was struggling through a very different sort of close encounter. The congratulatory press briefing had begun routinely and Beggs had said all the usual things about the Voyager triumph ("This nation, when it puts its mind to it, can do splended things.... It staggers the imagination . . . one of the great scientific achievements of our age.") Then Beggs sat down behind the big desk on the Von Karman stage and waited for questions from the press. Within ten minutes, he looked like a man who desperately wanted to be somewhere else.

For the first time in anyone's memory, pack journalism came to JPL. The space gypsies, always inquisitive but normally tame, pounced on Beggs, hitting him with a barrage of tough, specific questions about NASA policy and the future of American planetary exploration. Beggs, who had taken office less than two months earlier, dodged and weaved with his first few answers, but the journalists weren't buying it. The congratulatory briefing quickly turned into the most adversary, downright hostile press conference anyone had ever seen at JPL.

Beggs told the press, "I cannot conceive of NASA without a strong science program." Perhaps Beggs couldn't, but back in Washington the Reagan Administration's budget cutters were planning just such a NASA. The Galileo mission to Jupiter, for which $300 million had already been spent, was facing cancellation. The Venus Orbiting Imaging Radar mission seemed to be in limbo. The American half of the International Solar Polar Mission, a joint venture with the European Space Agency, had already been cancelled. The once-in-a-lifetime chance for a mission to Halley's Comet in 1986 was slipping away with each passing day. Beggs, mumbling his way through evasive non-answers, said that the Halley mission was still "under consideration"; but the word from Washington was that the Halley mission would not receive funding, and by year's end the comet probe was scrapped. Japanese, European, and combined French and Soviet teams would all fly Halley missions (with less sophisticated spacecraft), but the Americans would have to be content to wait until Halley's next appearance in the year 2062.

In their passion to turn supply-side economic theory into reality, the Reagan Administration had slashed the budgets of every federal agency except the Defense Department, but NASA was especially vulnerable. The space-science budget, in real dollars, had been declining ever since 1976. As Voyager Project Scientist Ed Stone put it, "At some point, you reach the critical minimum, below which you don't have a viable program." By August, 1981, that point had been reached.

Beggs and other NASA officials tried to put the best face on the situation, but their efforts were unconvincing. Some promised that by 1986, once the Space Shuttle had been paid for, more money would be available for space science. Those promises struck Acting Associate Administrator for Space Science Andrew Stofan as "very interesting, but science is not going to be there anymore if we don't solve the problem before 1986." Even Stofan's title seemed ominous; a year after Tim Mutch's death, no permanent Administrator for the Office of Space Science had been appointed. Reliable sources in

Washington reported that Budget Director David Stockman had said on several occasions that NASA would be completely out of space science by 1984.

In the year since their gloomy meeting in Tucson, the space scientists had at last begun to fight back, but it was an uphill struggle. In a "Dear Colleague" letter, Chairman David Morrison of the American Astronomical Society's Division of Planetary Sciences wrote: "The time has come to politicize the planetary science community. . . . I realize this is distasteful to many, and I am sure all of us would much rather pursue our science. But we believe the danger is real, and we are not crying 'wolf.'"

The danger was, indeed, real and growing. By the time the planetary scientists met again, in Pittsburgh in October, 1981, there was one more horrifying possibility under active consideration in Washington. The Administration was weighing the possibility of pulling the plug on Voyager 2. The Deep Space Network was expensive to operate, and with no more deep-space missions in the works, NASA was considering shutting it down. Voyager 2 would encounter Uranus on schedule in January of 1986, and even without further instructions from Earth, the spacecraft would conduct its science program and broadcast the new data. But back on Earth, no one would be listening.

At Pittsburgh, the scientists finally responded to the alarm that had been sounded a year earlier in Tucson. Rhetoric would not be enough—a specific plan of action was needed. The scientists spoke of organizing letter-writing campaigns, lobbying Congressmen, and even moved toward the formation of a political action committee. While a few scientists still clung to the notion that even in the face of professional extinction, they had to remain "purer than Caesar's wife," to most of those gathered in Pittsburgh, it was clear that for planetary science this was the Ides of March. Unless they were willing to play political hardball, Caesar's wife would soon be a widow.

The scientists were battling, not just the Reagan budget cutters, but their own political naiveté. In an impassioned speech, one scientist declared that their meeting would be a failure unless the next day's headlines proclaimed: "Scientists Vow to Fight Cuts." There was such a story, under a milder headline, in the Pittsburgh paper the next day—on page seven. The page-one stories that week were more concerned with the impending layoff of 600 steelworkers. The country was entering a recession, unemployment was skyrocketing, and cuts in federal social programs had left millions wondering how they were going to survive. The problems of 250 planetary scientists did not concern most Americans.

The scientists were paying the price for years of aloof complacency. They might hustle after grants and funding for their own research; but when it came to raising public support for space exploration, most of them were content to leave the field to Carl Sagan. At Tucson, they had joked about Sagan's activities; now, in Pittsburgh, they were desperately trying to emulate him.

Ironically, while the space scientists didn't know how to sell their own program to the American people, big businesses were using space science to sell their products. Both *Time* and *Newsweek* launched major advertising campaigns featuring their cover-story issues on Saturn. There was a good reason for this—space covers had consistently been among the best-selling issues of both magazines. RCA had used Voyager images of Jupiter to sell color television sets. The Kelly-Springfield Tire Company was even selling a Voyager radial tire ("Voyager is sending back fantastic pictures—from the *road!*").

Obviously, the American people were not apathetic about space exploration. They were uninformed. In a year when Americans had spent 32 billion quarters shooting down video space invaders, most of them were unaware that their own space program was on the verge of being shot down for the lack of a few hundred million dollars. NASA could run a viable, productive planetary exploration program for less than $250 million a year—less than the price of a single B-1 bomber. People simply assumed that the space program would continue, as it had throughout the entire lifetime of most of the population. Congress would appropriate billions for MX missiles and neutron bombs because they perceived an anxious demand for such programs. In the absence of tangible support for space, the legislators were likely to acquiesce to the Reagan Administration's attempt to gut NASA.

So James Beggs looked uncomfortable in the face of the hostile questioning by the JPL press corps. He was being called to account, and he had no answers to offer. If anyone was in a position to fight for the planetary program, it was NASA Administrator James Montgomery Beggs. But Beggs, a graduate of the Naval Academy, seemed more willing to go down with the ship than make the effort necessary to save it.

The end of the planetary program might not signal the Dark Age James Michener had invoked, but the storm clouds were gathering. In the long run, a ten- or fifteen-year hiatus in space exploration might seem inconsequential. But there was reason to doubt that the program could be put back together after such a prolonged standstill. The crop of promising young physicists, geologists, and astronomers now deciding their own futures would not choose planetary science if there were no funding for it. Surely a new generation of children with the scientific expertise necessary to resurrect the space program would not spring from the classrooms of the Moral Majority. Darwin had already been expunged from the textbooks of major publishers—would Copernicus be the next to go? Would the Voice of the Space Age be stilled forever?

These were the questions hanging over JPL as Voyager 2 sailed past Saturn. The space gypsies were already calling it "The Last Picture Show," and there was a very real possibility that it truly would be the last—ever.

It was to be a bittersweet encounter, a poignant blend of champagne and tears, memories and dreams. The space science community—scientists, engineers, journalists, dreamers, and doers—gathered once more in their high-tech Brigadoon: here today, gone tomorrow, perhaps never to return.

But, of course, there was still Saturn. . . .

"Welcome aboard," said Ed Stone at the first briefing on August 21. The spacecraft was healthy, and Saturn lay just ahead. The bewildering results from Voyager 1 had led to an extensive redesign of the Voyager 2 encounter sequences. There would be a spoke movie; close passes at Enceladus, Tethys, and Iapetus; high-resolution images of the F ring, and an exhaustive search for the "embedded moons" that the scientists thought (and hoped) were hiding in the rings.

In addition, Voyager 2 would have a functioning photopolarimeter. On Voyager 1, the instrument had failed completely. During the Voyager 2 Jupiter encounter, the photopolarimeter had been turned off for fear of a chain-reaction malfunction in the high-

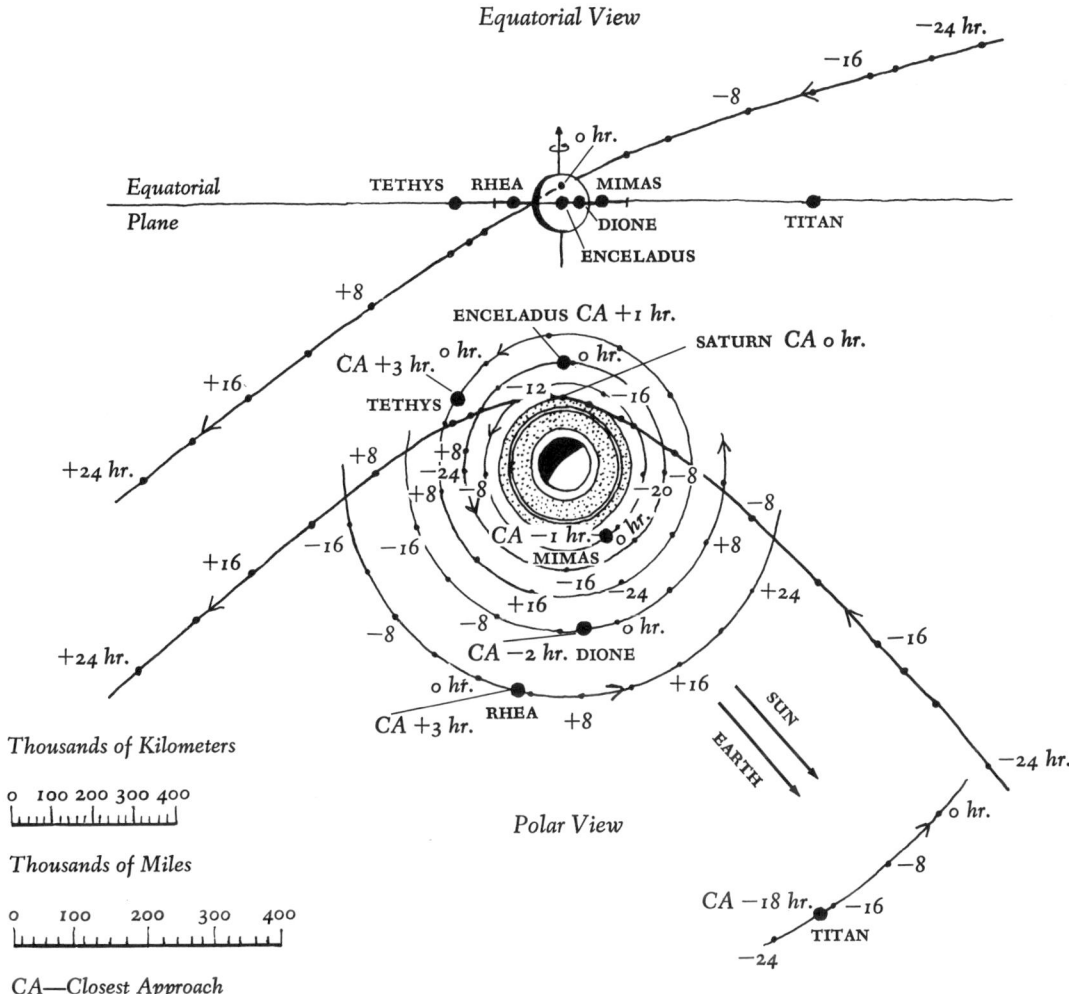

radiation environment. Now, after months of painstaking work, Arthur L. (Lonnie) Lane and his team had their instrument working again. With it, they would be able to conduct a fascinating and enormously revealing experiment. As the spacecraft passed over the rings on the in-bound leg of the encounter, the photopolarimeter would be focused on the star Delta Scorpii, below the rings. It would measure the intensity of the star's light a hundred times each second; as the star blinked on and off behind the rings, the photopolarimeter would, in effect, map the rings and gaps to a scale of a few hundred meters. The B ring, 16,000 miles across, would be seen down to the resolution of a city block.

As an unexpected bonus, it had turned out that Voyager 2's imaging system was much better than Voyager 1's. The spacecraft cameras had been chosen from a group of six similar but not identical instruments and, by the luck of the draw, the Voyager 2 narrow-

Voyager 2's trajectory included only one crossing of the ring plane, but it proved to be more than sufficient. Next stop: Uranus, in 1986.

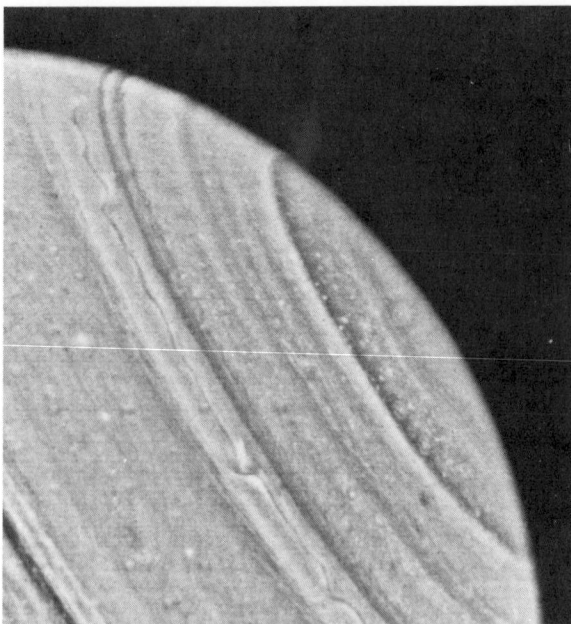
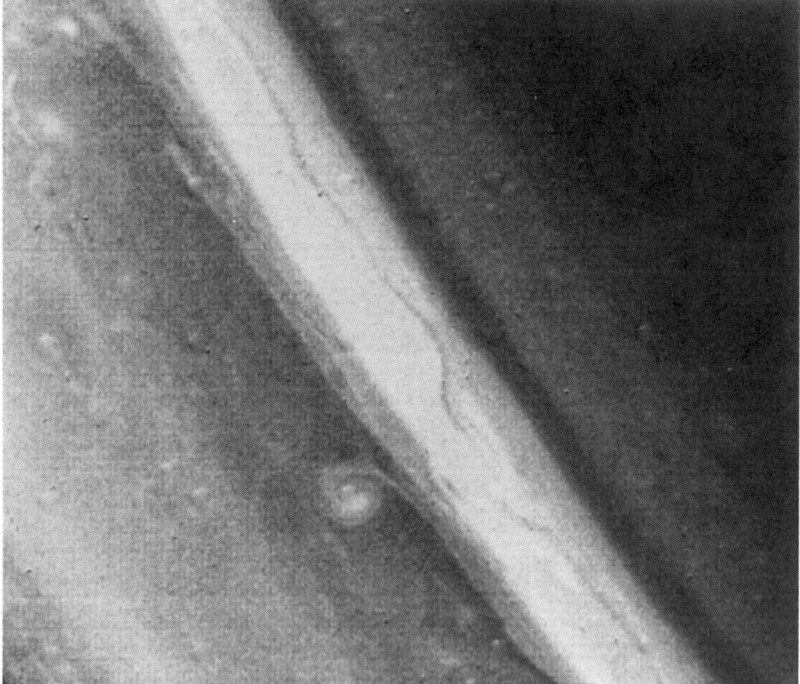

Computer image enhancement reveals the hidden structure of Saturn's atmosphere. Note the considerable activity at high latitudes and the wavy, ribbon-like high-speed jet streams. At left, a whirling cloud, similar to an Earthly hurricane, is about to break free from the zonal flow.

angle camera was fifty to one hundred percent more sensitive than Voyager 1's vidicon. The results were already visible on the monitors: Saturn had never looked better.

The planet itself hadn't changed a great deal. From Voyager 1 results, it had been determined that Saturn's general blandness, relative to Jupiter, was intrinsic. The hydrogen haze layer above the

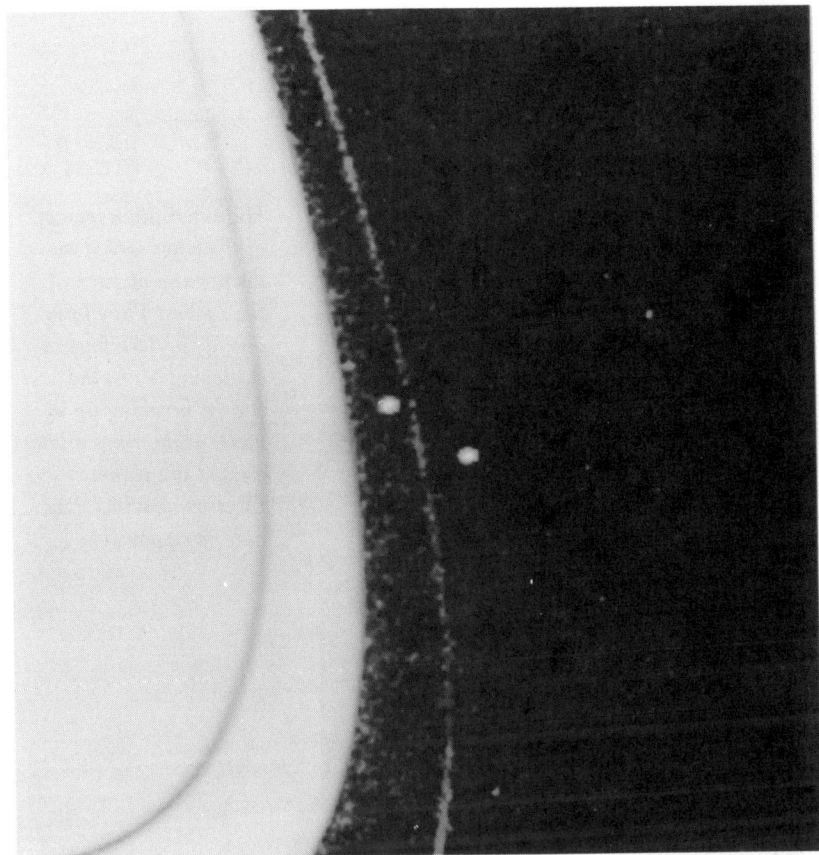

The two "shepherding satellites" chase each other around the mysterious F ring. The inner moon moves faster and laps its companion every 25 days. Scientists suspect that the two moons gravitationally focus the thin F ring and may account for its odd structure.

cloud tops, once thought to be responsible for the blandness, didn't contribute much to the muted appearance of the atmosphere. Apparently the chromophores (coloring agents) were simply mixing more thoroughly than they did in the Jovian atmosphere.

Voyager 2, thanks to its better cameras, was finding much more atmospheric detail than Voyager 1, particularly at high latitudes. "There's a feeling that the whole planet is more contrasty," said Brad Smith. The atmospheric bands were much more prominent. There was also a new large spot (10,000 km × 6,000 km) at 74°N, reminiscent of Anne's Spot in the southern hemisphere.

During the year since the Voyager 1 encounter, ground-based observers had found two more Saturnian moons, bringing the total to seventeen known satellites. The two new moons were barely 40 km in diameter and were stationed at the two stable Lagrangian points in the orbit of Tethys, sixty degrees ahead of and behind the major satellite. Voyager 2 would image both of them and look for others.

The two satellites shepherding the F ring had already been seen at long range. On August 15, the moonlets were in conjunction on opposite sides of the narrow ring, just 1,800 km apart. If the shep-

High-resolution images of the rings served to deepen the mystery of the spokes. They form quickly as dark fingers stretching across the B ring, breaking up in hours as the rings rotate around the planet. Electromagnetic forces may be responsible.

herds were causing the kinks and braids, the effect should have been seen in the conjunction images. But, Brad Smith reported, "We have to say that we see no effect whatsoever."

Ring viewing in general was better, mainly due to the more favorable sun angle (7°), but the better resolution so far had not led to any answers. The spokes were still there, still inexplicable. They had been seen from twice as far away this time, and a few high-resolution images seemed to have caught spokes in the process of forming.

The spoke movie followed specific spokes as they rotated with the rings all the way around to the shadow side. "It's almost like watching a car on a racetrack," said Rich Terrile. Inevitably, the movie became known as "The Saturn 500." The movie showed spokes forming along radial lines; the inner material then moved ahead, the outer portion of the spoke lagged behind, and the center of the spoke continued to rotate in synchrony with the center of Saturn. There was some suggestion that spokes might somehow be forming on top of each other; but the precise mechanism of spoke formation was still, as Terrile put it, "the $64,000 question." Photon particles from the sun may have been imparting an electric charge to the fine spoke material; this would be a way of segregating oppositely charged particles and might explain the electrical discharges heard by the planetary radio astronomy experiment.

Aside from the enduring scientific mysteries, the main focus of attention for this encounter was the precise location and shape of the G ring. Unlike Voyager 1, Voyager 2 would cross the ring plane just once, 54 minutes after closest approach to Saturn on the evening of August 25. The crossing point, behind the planet, would be about 2.7 Rs, a spot chosen years earlier because it was clear of visible rings and because a close approach was necessary for the gravity assist to Uranus. But now there was Pioneer's G ring, dangerously close to the crossing point. If all went well, the spacecraft would miss the G ring by about 1,200 km; but if the G ring turned out to be eccentric, like several others, there could be problems. A NASA Mission Operation Report summarized the situation: "The ring-plane crossing will be very close to the G ring, the eccentricity of which (if any) is unknown but, if real, could result in a ring penetration which would take place behind the planet; offsetting the potential concern is the knowledge that the ring is narrow and not dense, that the crossing angle is steep, and that there are no other alternatives if the spacecraft is to swing on toward Uranus—a vital scientific goal. In any event, the ring-plane crossing will be a point of some drama, heightened by the fact that it will take place while out of contact with the ground."

The first big event of the encounter was the closest approach to Iapetus on the evening of the 22nd. Voyager 2 would pass within 910,000 km of the two-tone moon, with maximum imaging resolution of about 17 km. The long-range images from Voyager 1 had simply deepened the Iapetus puzzle rather than solving it. "The fascinating thing," said Hal Masursky, "is that the contrast is so great." The dark material, with an incredibly low albedo of three percent, was more than ten times darker than the bright material. It was still possible that the dark material was coming from Phoebe, which was due to be imaged on the outbound leg. But the sharp dividing line between the bright and dark terrains suggested that internal processes were somehow responsible for the differing surfaces. "This is a very

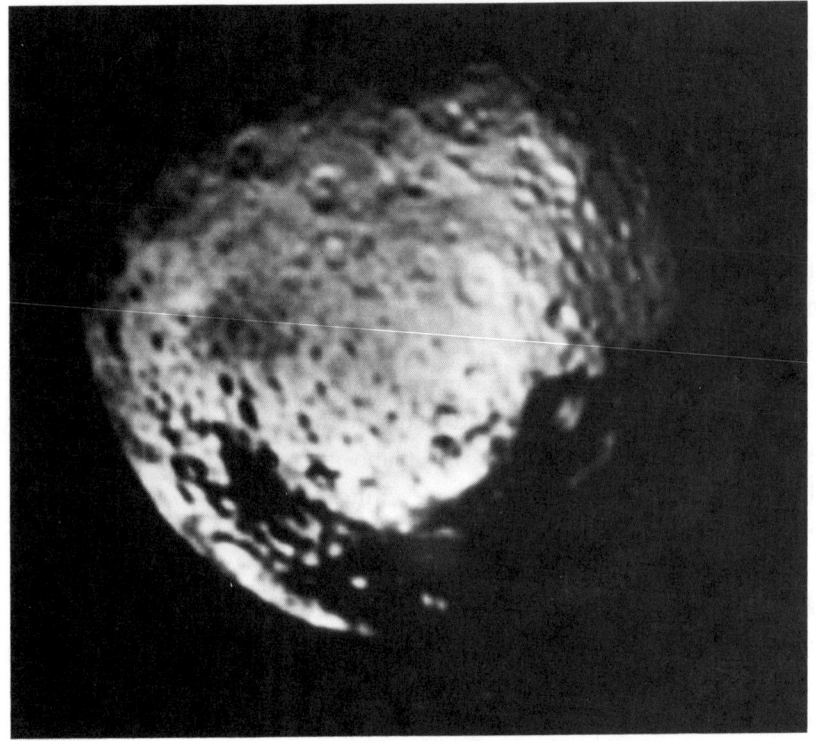

In an image processed to bring out details in the bright terrain, Iapetus reveals a heavily cratered, ancient surface. The bright material is probably ice; the dark material may be dust from a neighboring moon or an organic, asphaltlike substance.

strange, very exotic surface," said Masursky. He readily conceded that on Iapetus, as in the rings, "The effects we see are more complex than the ideas we've evolved so far."

One interesting idea about Iapetus was that the dark material was similar to the composition of a class of meteorites known as carbonaceous chondrites. The chondrites, which are thought to have originally formed in the asteroid belt, or possibly beyond, are relatively light, crumbly, and extremely dark. They are composed of organic, low-density, carbon-rich compounds. Carl Sagan, speculating on the nature of the dark material of Iapetus, said, "Organic is a good guess."

As it turned out, it may had been a very good guess. The close-approach pictures, in themselves, provided no definite answers. The crucial bit of knowledge to be gained from the encounter was the precise mass and density of Iapetus. That would be determined following close approach by a careful analysis of the spacecraft's trajectory. The denser Iapetus was, the more its gravity would bend Voyager's flight path. The best pre-encounter estimate was that Iapetus' density was less than 1.5 g/cm^3. A density of about 1.3 seemed most probable and would be consistent with solar composition models. If the density was lower still, then Sagan's organics would be a strong candidate for the dark material.

The pictures, meanwhile, were exciting ("This is it," Dave Morrison said; "this is the highest-resolution picture we're going to get in our lifetimes, probably.") and contradictory. Iapetus was heavily cratered, yet the craters themselves didn't quite seem to fit any model. If the dark material was thin, the impacts ought to have excavated bright material, creating bright rays. But there were no bright-rayed craters in the dark terrain, although there were a few anomalous bright spots. On the other side of the moon, there were no dark-rayed craters in the bright terrain. These inconsistencies made interpretation difficult. After Morrison discussed Iapetus on the Blue Room TV circuit, Al Hibbs said, "We'll talk to you again when the mystery begins to deepen or go away." "It will probably do both simultaneously," Morrison responded.

Predictably, it did. By the morning of the 23rd, the scientists were studying the Iapetus images, looking for clues that would favor one model over the other. Brad Smith noted that the dark material was three or four times darker than the black basalt flows on Earth's moon, adding, "That still doesn't tell us where it came from." If the dark material was exogenic (from Phoebe) it ought to have fallen over the surface of Iapetus like "frozen sunlight." In a given crater, in-falling dark material ought to have accumulated along one crater wall, leaving bright material on the opposite face (in the "shadow" of the Phoebe dust). So far, there was no evidence that this had happened.

Though the other Saturnian satellites were somewhat less mysterious than Iapetus, the system, in toto, was still a confusing mess. "It's a very junky system," said Torrence Johnson. "The place just seems to be full of chunks of ice everywhere we look." The ice was mixed, almost randomly, with chunks of rock. "Even the regularities are irregular," said Masursky.

One slight hint of regularity was seen in the fact that there were no rocky moons inside the orbit of Rhea. The inner bodies were almost entirely ice and were all quite small; the closer they were to Saturn, the smaller they were. That implied that the collision rate close to Saturn during the early days of the solar system was so high as to be lethal to any rocky moon. "If you had a parent body for S-10 and S-11 [the co-orbitals]," said geologist Gene Shoemaker, "you can predict it would have been broken." The lifetime of any small body close to Saturn had to be quite short. From there, Shoemaker argued, "It's not a very large step to say that the ring material may be the product of collisions." However, the total mass of the rings was equal to that of a satellite just 200 km in diameter. If the rings were mashed-up moons, the parent bodies must have been tiny.

All the satellite surfaces seen so far were clearly ancient. Although there was evidence of some internal geological activity left on the surfaces of a few of the moons, cratering rates indicated that even that must have taken place several billion years ago. The only exception was Enceladus, still to be seen at high resolution. "I think we've all got our eyes on Enceladus," said Larry Soderblom. "It's a very screwy body. I think we're all anticipating that we may be in for another bizarre and geologically active planet."

At the background session on the satellites, the geologists kicked around the prevailing theories about how the Saturnian system formed. At this point, no one seemed to have a great deal of confidence in any of the existing theories. Kelly Beatty asked if anyone had a "dark horse candidate," and—rather surprisingly—someone did.

The dark horse belonged to Gene Shoemaker. He was quick to point out that the theory was "so new that it's not even published yet." Highly animated and fairly bursting with enthusiasm, Shoemaker outlined the essential points of his theory.

Looking at the disarray in the Saturnian system, Shoemaker saw evidence that there had been "an *incredibly* intense period of bombardment at Saturn, sufficient to have smashed up *totally* any pre-existing satellite system. Then, you build them up all over again." Massive Titan might be the lone survivor of the blitz; the other moons would all be younger, built up over a half billion years from the scraps of the original system.

All this devastation, in Shoemaker's theory, would have been produced by comets. The nuclei of comets tend to be rather small—only a few tens of kilometers across—but early in the history of the solar system the typical comet nucleus may have been more massive. One surviving example may be Chiron, the large asteroidlike body discovered in deep space in the seventies by Charles Kowal. Chiron may be representative of the class of bodies that smashed into the proto-Saturn system. According to Shoemaker, there could have been an enormous number of such bodies, perhaps a total mass some ten times greater than the mass of Jupiter.

Astronomers believe that today nearly all of the comets orbiting the sun reside in what is known as the Oort cloud, a distant 40,000 AU away. It is unlikely that they formed out there; they must have begun life among the planets where more building material was available. Then, as the planets formed, their gravitational fields distorted the orbits of the comets, not unlike the manner in which Jupiter and Saturn altered the trajectories of the Voyagers, ultimately ejecting most of them from the inner solar system. But as they left, they clobbered Saturn's original moons. The new moons formed from the leftover junk, and on their surfaces we see, according to Shoemaker, "a record of the last stages of the bombardment."

The theory could account for such oddities as the huge Death Star crater on Mimas. It also seemed to explain the irregularity of the Saturnian system. After the original smashup, there would be a well-mixed collection of rock and ice orbiting Saturn. As the new moons accrete from the junk, they are heated by the final stages of the bombardment. The heating would melt the ice, and the rocky material would sink to the core. Over all, in rebuilding the moons, one would expect an essentially random distribution of small bodies with differing compositions and densities—which is almost precisely what we see at Saturn.

Shoemaker's theory is new and may not survive for long. However, subsequent discoveries by Voyager 2 did nothing to discredit it and, in some ways, tended to support it. Whether or not the theory holds up, it is an excellent example of the intellectual ferment created by planetary encounters.

As new theories were born, old ones died, and a few hovered on the brink of extinction. Andy Ingersoll's nested-cylinder theory of the interior of gas giants had not fared well in the ten months since the Voyager 1 encounter. "I'm the only one who's still pushing it," Ingersoll conceded at a background session. His colleagues, Garry Hunt, Reta Beebe, and Vern Suomi, offered little support. "I think Andy's model requires great symmetry between the northern and southern hemispheres," said Suomi. There was a gross similarity between hemispheres in both Jupiter and Saturn, but the cylinders needed something approaching mirror-image symmetry.

DISTANT ENCOUNTERS

Unfortunately, on both planets the search for symmetry was complicated by anomalies. On Jupiter, said Suomi, "The Red Spot screws it up." On Saturn, atmospheric dynamics were influenced by the cold shadow of the rings, which kept sunlight from reaching near-equatorial regions for years at a time.

Aside from symmetry, the convective regions in the gas giant atmospheres seemed to be at odds with the cylinder theory. On both planets, the regions of hot, rising gases were near latitudes where wind speed relative to the core was close to zero. To Beebe, this implied "a direct line of communication to the core." Ingersoll's rotating cylinders would smear that direct line.

But Ingersoll wasn't ready to give up. The cylinders worked well in explaining the oval-shaped features in the atmosphere. The ovals always maintained the same latitude and seemed to form at the boundaries between the hypothesized east and west spinning cylinders. The oppositely directed winds confine the ovals to the latitude where they form. Computer predictions based on the cylinder model conformed well with what Voyager actually saw.

To resolve the controversy, the imaging scientists were concentrating on the atmospheric ovals. By carefully plotting the wind speeds and directions in the ovals, they might find the symmetry

Forecasting the weather on Saturn depended on careful tracking of atmospheric spots and ovals. By following their movements, scientists hoped to gain an understanding of the unseen interior beneath the cloudtops. The large spot in the photo at bottom right is 1,900 miles in diameter and moves eastward at about 65 mph.

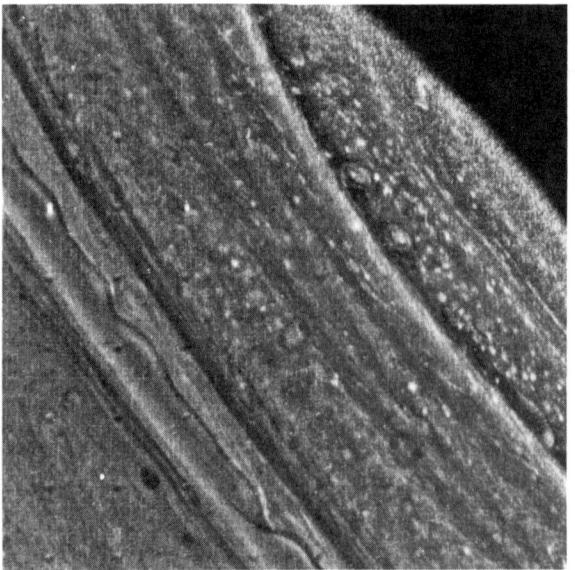
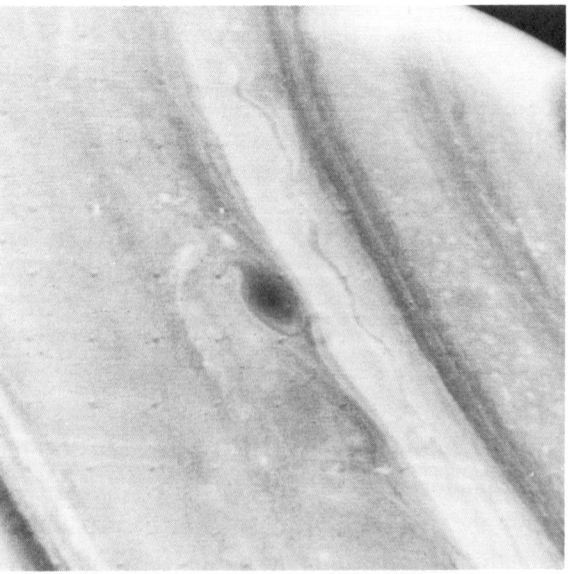

234

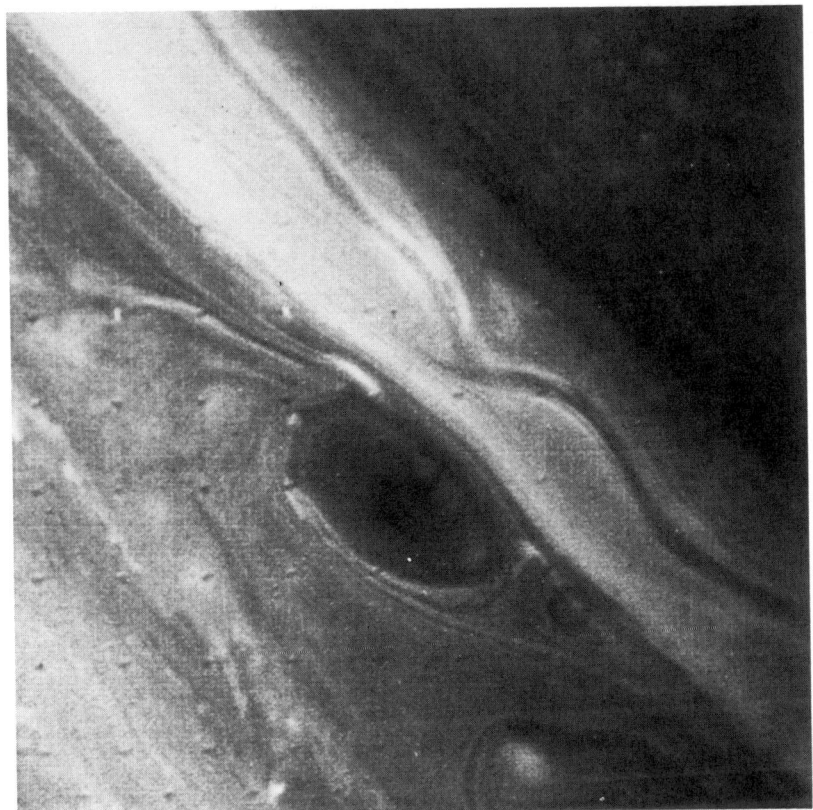

The counterclockwise (anticyclonic) rotation of this northern hemisphere spot may be associated with downdrafts, which would induce clearing over the spot. Thus the spot may be an opening into the deep atmosphere beneath the cloudtops.

Ingersoll needed or, conversely, an unambiguous example of a nonsymmetrical feature. They had planned a sequence of high-resolution images of Anne's Spot ("We're hoping," said Ingersoll, "that by the time the IAU meets, these spots will have disappeared so we won't have to name them all."), but there was no guarantee of success. To get images of other spots, they had to know where to aim the cameras. "We have no idea if it's going to work," said Ingersoll. "It's six-week weather forecasting on Saturn."

Two days later, the narrow-angle images of Anne's Spot arrived, showing an enormous amount of detail. "Andy Ingersoll did a superb job of long-range weather forecasting," said Brad Smith. Now, the task facing the scientists was the detailed analysis of the images, a process which would take years. After seven encounters with Jupiter and Saturn, we were finally at a point where it might be possible to get some sort of handle on the nature and behavior of gas giants.

The rings were another matter entirely. The search for the embedded moons was not going well. An icy moon of 10 km or more would have been readily apparent, but so far—nothing. On the 23rd, Brad Smith reported that the search program was one-third com-

The closer Voyager looked, the more rings it saw. In these images of the A (upper left), B and C (above), and B rings (left), hundreds of ringlets are visible. The smallest features visible here are some 10 miles wide, but the photopolarimeter experiment found distinct ringlets no wider than a city block. Variations in brightness are caused by differing particle size and density. Wave phenomena may be responsible for the complex structure of the rings.

pleted. "We have looked very, very carefully," particularly in the region of the C ring and the Cassini Division. "If we don't find those satellites, I think we're in real trouble."

By the next day, it was becoming obvious that the embedded moons simply didn't exist. Said Smith: "We now find ourselves at a point where we had hoped not to be. We are looking—desperately now—for other solutions." The ring search program was not complete, but based on the results from the Cassini Division, there was little reason to think that the moons were going to turn up anywhere. "For the most part," said Smith, "at least for the Cassini Division, we have to rule out the possibility that these two gaps were caused by an embedded satellite."

The demise of the embedded moonlets didn't help those trying to explain the braids in the F ring. Since the shepherding satellites didn't seem to be having any apparent effect, perhaps even smaller moonlets were involved—shepherds for shepherds. The mini-shepherds could be controlling the three strands seen in the Voyager 1 images of the F ring. Two strands would wiggle and one wouldn't, producing the braided appearance. But the small size of the F ring's particles seemed to imply that nongravitational forces were at work, and the mini-shepherds might not be necessary. What *was* necessary was more data.

The ring analysts were being pushed reluctantly in the direction of wave phenomena as the only plausible explanation for the overall structure of the rings. Inherent instabilities in a flat, spinning disk could create what Larry Esposito of the University of Colorado referred to as a train of waves rippling across the disk, with the wavelength decreasing toward the outer edge. This terribly complicated theory was currently in vogue as an explanation for the structure of other large, flat, spinning disks—the spiral galaxies. But spiral galaxies are immense, tens of thousands of light-years in diameter. Could a phenomenon at work in such huge structures apply equally well to the tiny particles in Saturn's rings? No one could say, but in the absence of the embedded moons the wave theory was emerging as the only plausible alternative.

There was also structure in the rings that was not easily explained by either moons or waves. False color images, keyed to reveal different particle sizes, showed a definite difference between particles in the C and B rings. That was consistent with Len Tyler's radio occultation data, which revealed distinct populations of different-sized particles ranging from centimeters in diameter to a maximum of about 12 meters. However, the false color images also showed some surprises. "The thing that intrigues me," said Brad

Smith, "is that we see two or three features in the C ring that have the same color as the B ring." Whatever force was controlling the ring structure, it seemed to be sorting out the particles in some pattern that was neither random nor obviously logical. What were B-ring particles doing in C-ring ringlets?

Meanwhile, the spokes were continuing to cause headaches. Rich Terrile said it was "very tempting" to try to tie the spokes to the lightninglike electrical discharges heard by the Voyager 1 PRA experiment. The source region was "localizing in the right area," which happened to be the thickest part of the B ring, home of the spokes. However, the discharges were not easy to link to the spokes. The discharges seemed to be extremely localized; the spokes were more than 10,000 km long, but the source region for the discharges seemed to be no longer than a few tens of kilometers. If the spokes and discharges were truly related, the precise mechanisms involved were still a mystery.

Saturn's periodic radio emissions still made little sense to the particles and fields men. The periodicity implied that the emissions were tied to some irregularity in Saturn's magnetic field. Yet Saturn's field was depressingly regular, symmetrical to within one degree. Axially symmetrical magnetic fields were thought to be impossible, but there it was. One way out was to postulate field irregularities in the deep interior of the planet where they can't be measured directly. The conducting shell of metallic hydrogen surrounding the core could somehow damp out the irregularities before they became apparent at the surface. The only problem was that there was no evidence to support such an idea. "If you'd like to appeal to an Act of God clause in your scientific theories," said experimenter Norm Ness, "feel free to do so."

The radio emissions were not the only broadcasts coming from Saturn. At the briefing on August 25, Fred Scarf played a unique tape of sounds generated by Titan's magnetic wake as Voyager 1 passed through it. The journalists enjoyed the performance, partly because it required no deep understanding of abstruse physics or other mental gymnastics. Scarf fed his data through a specially programmed Apple II home computer tied to a music synthesizer. The result was an eerie, haunting series of electronic chirps, whines, and hums that sounded like background music for a science-fiction film.

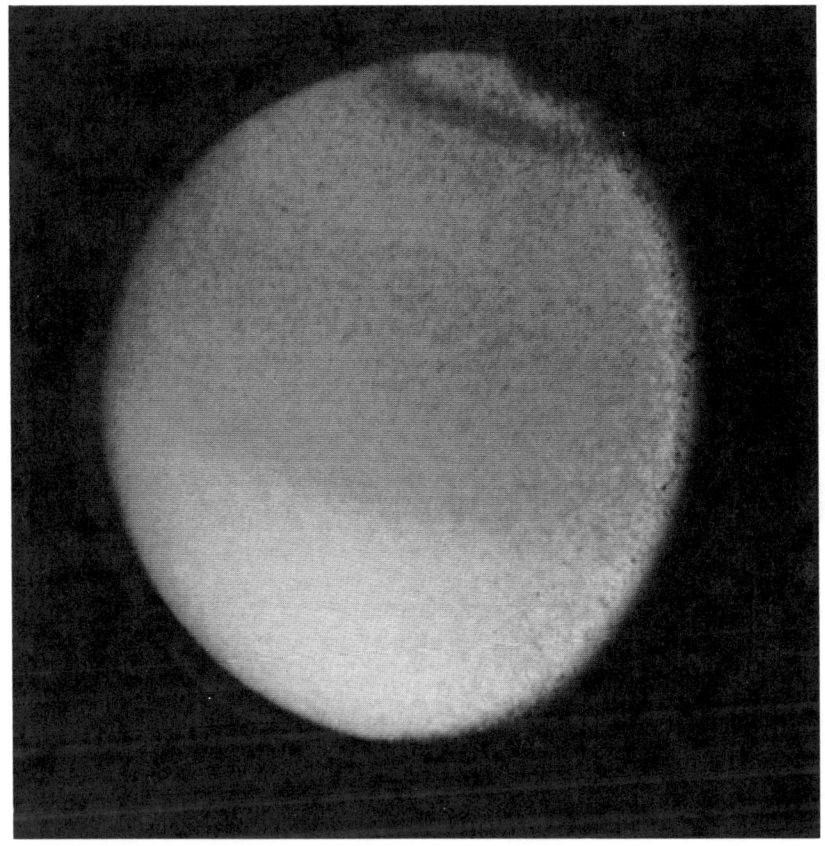

Titan's bland clouds reveal hints of structure in this image taken through a violet filter. Note the polar "hood" at about 60° north and the sharp equatorial band dividing the dark north from the brighter south. Beneath the clouds, organic molecules may be raining down into a sea of liquid methane.

Titan might be playing a different tune this time around. Voyager 1 had seen it inside Saturn's magnetosphere, but now it was on the outside. Voyager 2 hit the bowshock on the afternoon of the 24th and made five crossings before entering the magnetosphere at 18.6 R_s. Instead of being bombarded by trapped radiation, Titan's upper atmosphere was now being hit by solar wind particles. The change was probably not related to a subtle alteration in the appearance of Titan. Brad Smith said it was still "a rather uninspiring orange ball," though the north polar hood now looked less like a hood and more like another atmospheric band similar to the equatorial band already seen. Some scientists suggested that this was a seasonal effect, although it had only been one Saturnian "week" since the Voyager 1 encounter.

The other satellites were coming into view now, and the imaging had been routinely spectacular. Hyperion, a tiny outer moon orbiting between Titan and Iapetus, was an odd-looking hunk of rock that was variously described as a peanut, a hamburger patty, and a hockey puck. The fascinating thing about it was that its major axis was not pointed at Saturn, suggesting that the moon might not be in

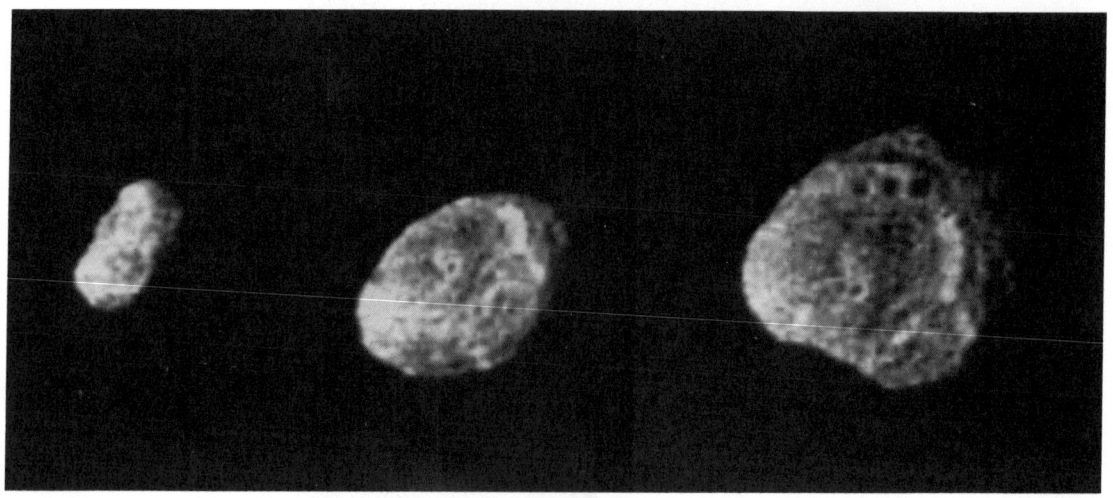

tidal lock with the planet. It was possible, in fact, that it was in some sort of tidal lock with Titan. "It's not tumbling wildly," said Smith, but for it to be even slightly out of lock was significant. One possible cause: a very large impact might have knocked the moon out of alignment. In any case, Hyperion was interesting because, if the dark side of Iapetus was sweeping up Phoebe dust, some of the dust ought to have hit Hyperion as well. Its color was reddish, dark enough in places to suggest that it was collecting dust; yet its overall albedo was a relatively high thirty percent.

Three views of Hyperion (top) reveal a pocked, hamburger-shaped moon, 220 by 130 miles. The tiny moon may have been knocked out of a gravitationally stable orbit by a large impact. A huge impact was responsible for another "Death Star" crater, this one on Tethys (left). The 250-mile crater is visible near the day–night line and is on the opposite side of the planet from the great trench seen by Voyager 1.

240

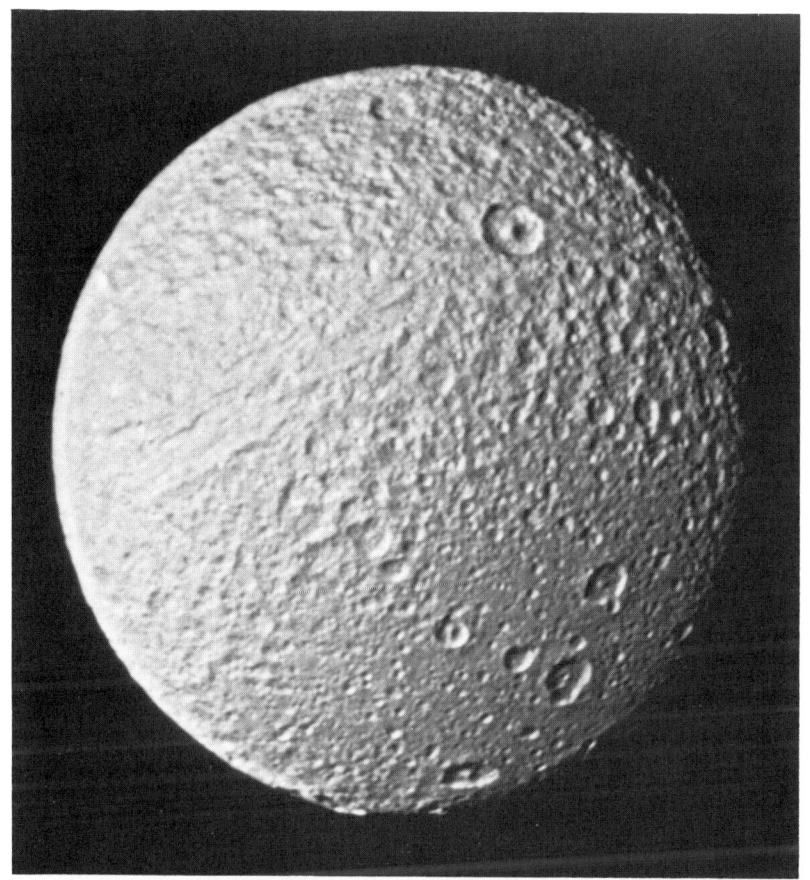

Computer processing brings out the details of the surface of Tethys. Note the contrast between the heavily cratered terrain at top right and the less battered terrain at lower right. This may be evidence of internal geologic activity. The great trench, at left center, runs three-quarters of the way around the moon.

This surviving fragment of the lost high-resolution mosaic of Tethys reveals details as small as 1.4 miles across. The surface of the 650-mile-diameter moon is mainly ice.

241

By the 25th, the mass of Iapetus had been calculated. The moon weighed in at about 1.1 grams per cubic centimeter. That made it considerably less dense than predicted, and limited its possible rocky mass to about twenty percent. Sagan's model of some sort of organic, carbonaceous material seemed quite likely now.

Voyager 2 saw the opposite face of Tethys for the first time and revealed that the huge trench seen in Voyager 1 images extended for three-quarters of the way around the planet. But the most eye-catching feature was another mammoth Death Star crater, more evidence for Shoemaker's bombardment theory. Relative to the size of the planet, it was roughly the same size as the crater on Mimas; but Tethys is a much larger moon, and Mimas itself would fit comfortably inside the 400-km crater on Tethys. At first, it looked as if the crater was located precisely opposite the trench, implying that the impact had cracked the planet. It turned out, however, that the crater was a bit off-center and probably predated the trench. The trench, incidentally, was now known as Ithaca Chasma. Despite what some may have thought, the name was not a tribute to astronomers from Cornell, which is located in Ithaca, New York. Rather, it is consistent with the naming scheme adopted for the Saturnian satellites, based on place-names in *The Odyssey*.

The satellite everyone was waiting for was Enceladus. The first good pictures were due on the evening of the 25th, shortly before closest approach to Saturn. Following ring-plane crossing and occultation, plans called for additional high-resolution coverage.

August 25 turned out to be one of the more eventful days in the history of planetary exploration. JPL was mobbed, with the largest press corps of the entire mission in attendance. It was a bad day to be part of a large crowd, however; the temperature in Pasadena was a sweltering 106°F, and there was a smog alert. The weather may have contributed to the testy confrontation that morning between the press and NASA Administrator Beggs, but by evening things had cooled off, meteorologically and emotionally.

The Enceladus pictures were especially poignant for veterans of the Voyager encounters. Enceladus would be the last moon to pose for Voyager's high-resolution cameras until 1986, the last scrap of new turf this side of Uranus. The scientists and space gypsies gathered in front of the monitors for one last session of Saturnian sight-

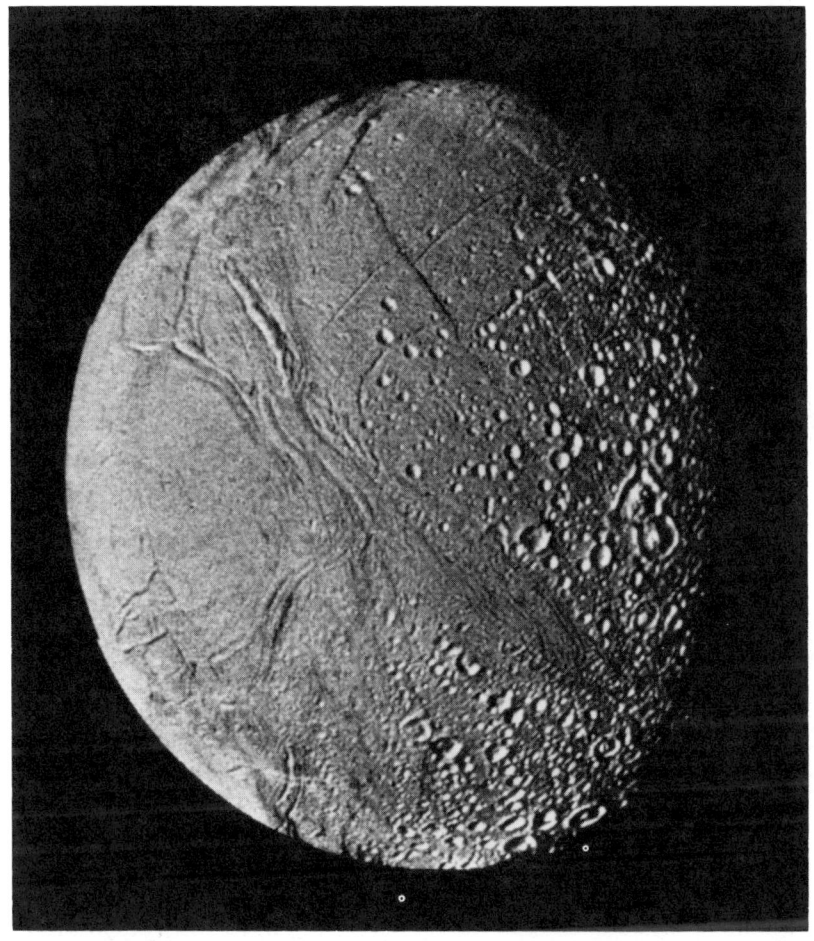

At least five different types of terrain are visible on Enceladus. In some regions, impact craters are almost completely absent, implying extreme youth and, almost certainly, a still-active geology. Although the icy moon in some respects resembles Ganymede, the Jovian satellite is ten times larger than Enceladus.

seeing. Although no one realized it, this was very nearly the final session, period. Voyager 2 was about to play the lead role in a nail-biting interplanetary melodrama.

Enceladus more than lived up to everyone's hopes. This was the Patchwork Planet, a crazy quilt of at least five different terrains; it looked as if it had been assembled from leftover pieces of Mercury, Mars, Ganymede, and Europa. Regions of shoulder-to-shoulder ancient craters were adjacent to blocks of complex twists and folds, reminiscent of the grooved terrain on Ganymede. There were huge, straight-line faults, almost geometrically precise. And meandering across the center of the disk was something that looked like a vast dry wash, where water may have flowed at some time in the past. In this region, there were virtually no craters at all. It had to be extremely young, no more than about 100 million years old. "There is no reason to believe this is only ancient activity," said Hal Masursky. In all probability, Enceladus was still geologically active.

Throughout the early evening hours, the spacecraft descended

Something strange was happening as Voyager 2 took these images of the rings, just minutes before ring plane crossing. These images emphasize the thinness of the rings, but they are even thinner than they look—perhaps no more than 100 yards thick.

toward the ring plane. As it did so, the photopolarimeter locked onto the star Delta Scorpii and followed it as it twinkled across the rings. When the signal confirming the operation of the instrument arrived, the photopolarimetry team cheered and whooped like a bunch of kids set loose in a toy store. The PPS team had been battling their cantankerous instrument for four years, and now it was finally working. Lonnie Lane passed out PPS buttons—Perfectly Phantastic Science.

It was, indeed, "phantastic." For two hours and twenty minutes, the strip chart in the PPS office cranked out dips and wiggles in the intensity of the light from Delta Scorpii as it was eclipsed by the rings. The experiment would produce a graph 2,700 feet long, with 25 data points per inch. The rings would be resolved down to 100 meters, far beyond the resolution of the narrow-angle camera. Even Brad Smith, who had to give up two and a half hours' worth of imaging for the sake of PPS, was impressed. Staring at the unrolling strip chart, he told Lane, "It was worth giving up the time." By the time Voyager's signal was lost behind Saturn, at 10:33 PM, it was clear that the number of rings was about to increase by at least a factor of ten. Lane and his team left the office at about 11:00, exhausted and ecstatic.

The spacecraft, meanwhile, was out of contact with Earth. While Voyager was behind Saturn, it would cross the ring plane. No one would know if it had survived its close brush with the G ring until midnight, when the signal was due to be reacquired. The accuracy of the trajectory was as good as anyone could hope for—after four years in space, Voyager 2 was arriving at Saturn just 3.1 seconds early. Charlie Kohlhase compared it to a golfer sinking a 500-mile putt (with the help of nine course corrections along the way). Voyager was within 41 miles of where it was supposed to be; the question gnawing at everyone was whether the G ring was where *it* was supposed to be.

A few seconds before midnight, Al Hibbs appeared on the Blue Room circuit. He was on the air less than ten seconds when the news arrived: "We just received the information that the signal has been reacquired!"

In Von Karman, the space gypsies cheered; they now knew what they would be doing on January 24, 1986. Cleverly designed buttons which read, "Goodbye, Saturn!" were turned upside down; in the new orientation, the same script now read, "Hello, Uranus!" Champagne corks popped, Uranus was toasted, and a good time was had by all. Tired and happy, the scientists and reporters went home for the night.

It was to be a short night. Within hours, scientists were being roused from their sleep by urgent phone calls from the Lab. Something had gone wrong. Voyager was taking pictures of empty space.

Shortly after midnight, it was learned that the scan platform was stuck in the azimuth, or horizontal, position at 260°. "That was not where it was expected to be at that time," Mission Director Dick Laeser said at a briefing early the next morning. "Within minutes of the problem, we had all the right people in here to work on it." At 2:00 AM, commands were sent up to the spacecraft which would point the scan platform away from the sun. The instruments on the platform (both cameras, IRIS, UVS, and the photopolarimeter) could be ruined if they were accidentally aimed too close to the sun. Until the glitch could be fixed, caution was the watchword.

Until now, the scientists had held center stage, while Project Manager Ek Davis (who followed John Casani, Robert Parks, and Ray Heacock in the top job) and Mission Director Dick Laeser served as spear carriers. But circumstances had conspired to make the managers the eleventh-hour heroes of the Voyager drama. In the days that followed, Davis and Laeser did a magnificent job, keeping the spacecraft alive and the press well informed. They worked killing hours, along with the hastily assembled spacecraft anomaly team, and projected a rock-solid competence that was reassuring even in the face of one of the most baffling events in the history of space exploration. Looking drawn and weary, Davis somehow reminded one of a pro quarterback leading his team out of the huddle late in the fourth quarter with the game on the line. Calm, cool, and confident, Davis was almost an American archetype, a distillation of fabled Yankee ingenuity, American knowhow, and last-ditch, remember-the-Alamo refusal to accept defeat. No less than Tom Wolfe's astronauts, these men had the right stuff.

At stake were the Uranus and Neptune encounters—in other words, the entire future (such as it was) of American planetary exploration. With the scan platform stuck, the only way to point the instruments on the platform was to roll the entire spacecraft. Charlie

Kohlhase estimated that "We could get 85 percent of the science for the Uranus encounter, assuming the only problem is the stuck axis." The limiting factor was fuel. Voyager 2 had a margin of 15 pounds of hydrazine propellant after all the course-change maneuvers needed to get to Neptune. A single roll of the spacecraft consumed $\frac{1}{10}$ pound of fuel. Thus, Voyager could do 150 rolls in the next 8 years and acquire a substantial number of images, assuming nothing else went wrong.

Nobody was even very sure exactly what had gone wrong already. The first indication was that the platform had stuck some 45 minutes after ring-plane crossing, but there was some uncertainty about the time. "We would like to think it was later, rather than sooner," said Ed Stone. The sooner it had happened, the more data would have been lost.

It was sooner. As the engineering telemetry was analyzed, it became clear that the scan platform pointing was slightly off even before the ring-plane crossing. There were also other indications that something bizarre had happened. According to Laeser, the UVS, PPS, and magnetometer sent "strange data for 10 to 15 minutes after acquisition, and then it cleared up." No one could say if the "strange data" were related to the stuck scan platform; the UVS and PPS were mounted on the platform, but the magnetometer was at the end of a 15-meter boom at the opposite end of the spacecraft.

Throughout the morning, the spacecraft replayed data acquired during occultation. Monitors throughout the Lab displayed pictures labeled "Saturn" or "Enceladus," but each image was the same: empty blackness. One of the two high-resolution photomosaics of Enceladus had been lost, along with a mosaic of Tethys. The missing Tethys pictures included narrow-angle shots of the trench with a resolution of just 1.7 km. The images might well have been among the most spectacular of the entire mission; now they were irretrievably lost. Also gone were sequences aimed at Saturn's southern hemisphere and the rings on the night side.

"We were fortunate that it [the platform] didn't stop a few hours earlier," said Ed Stone. "We've completed most, if not all, of the major science objectives." In the days that followed, Stone stoutly maintained that Voyager 2 had achieved "200 percent" of its planned objectives. When pressed, he would admit that without the glitch the total might have been 202 percent. The only remaining science target was the outer moon, Phoebe, which was due to be imaged on September 4. Because of its possible relationship to Iapetus, scientists were eager to get at least a few long-range images of the tiny satellite. Phoebe was also interesting because there was a chance that it was either a captured asteroid or an extinct comet nucleus; neither class of object had ever been seen from a spacecraft. But if the scan platform could not be fixed quickly, the Phoebe images would probably be lost.

The platform was going to be difficult to fix, it appeared, because no one had a clear idea of what had gone wrong. The obvious conclusion was that the spacecraft had sustained a damaging impact during ring-plane crossing. But the pointing problem had started before the crossing, when the spacecraft was thousands of kilometers above the ring plane in supposedly empty space.

The data, meanwhile, were yielding more indications that something truly weird had happened. The onboard computer was programmed to fire the spacecraft's thrusters whenever Voyager drifted off course. Shortly after the crossing, the thrusters fired 86 times. "That would imply some force being exerted on the spacecraft," said Ek Davis, "something hitting it—but after ring-plane crossing." The signals from the plasma-detection experiment were even stranger. "The data just went crazy," reported Davis. Again, it seemed to indicate that Voyager had collided with something, possibly a cloud of ionized dust.

Despite plenty of symptoms, the disease defied diagnosis. "I just don't know about causes," said Davis. The cause could be in the scan platform itself, or in the platform's gearbox, or in the computer. "It could even be in the software," said Davis.

By the morning of the 27th, the situation seemed to be improving slightly. The receiver was being troublesome again, and just keeping in touch with the spacecraft was a major concern, but after three tries, the controllers managed to get a new computer "load" into the spacecraft. With the program in, the spacecraft responded well to its new instructions.

"The question," said Ek Davis, "was can we move it at all? Last evening we commanded it to move, and it did. So that was good news. The platform did not always go at the rate we specified. It seemed to slow down a bit, then it speeded up to its normal rate." The success of the test sequence was encouraging, but the engineers were no closer to understanding what had gone wrong.

What made the anomaly so curious was the fact that Voyager 1 had experienced an almost identical glitch in the scan platform early in 1978 while it was still on the way to Jupiter. After an exhaustive analysis and many simulations, the controllers found the source of the problem. It seemed virtually certain that a tiny scrap of plastic in the gearbox had broken off and gotten jammed in the clockworks. By ordering the platform through a carefully planned series of maneuvers, or slews, the plastic was either crushed or brushed aside, and Voyager 1 was restored to normal operations. The Voyager 2 anomaly seemed remarkably similar, except for the additional weird data from the other instruments and the unexplained thruster firings. And if Voyager 2 was the victim of another scrap of plastic in the gears, why—after four years in space and a billion miles—*why* did it happen at the hour of ring-plane crossing?

On the morning of the 28th, there were still no answers, but there was reason for optimism. "There's a good chance that Saturn will be on our TV screens by the end of the day," Dick Laeser announced. "The stickiness and the slower than normal performance we've seen has gotten better . . . the platform gets better with use. . . . The last test seemed to be almost right on." With the situation improving, the controllers hoped to get the Phoebe observations on September 4. Phoebe would be a little too close to the sun to trust the scan platform on its own so the entire spacecraft was rolled to the correct position.

The crucial test sequence was scheduled for late afternoon. The platform was commanded to aim at the dark side of Saturn and resume imaging. At 5:41 PM, the first picture arrived. A faint, ghostly, static-ridden image appeared on the monitors. It was Saturn, dead center. The aiming was on the nose. But why was the picture so terrible? Had the cameras somehow been pointed at the sun? Did we now have a healthy scan platform and a diseased imaging system?

A few minutes after the image appeared, Dick Laeser came into the press room, looking happily bewildered. "Don't ask me to explain it," he said to the reporters who clustered around him. He pointed to a hand-drawn graph, plotting the platform's position versus time.

"You couldn't be much more right on," Laeser said. "That's where we wanted to be at that time, and that's where we are."

The platform, it seemed, was fixed. And the bad image? "Someone from the imaging team ran downstairs, practically in tears, bless her heart," said Laeser. The emissary from imaging explained that the picture was a simple mistake, an accidental underexposure. There was nothing wrong with the camera.

"There was dead silence when that picture came back," said Laeser. But now, all seemed to be well. "I've got my fingers uncrossed," said Laeser, demonstrating by crossing all his fingers simultaneously; if there had been any wood nearby, he probably would have knocked on it three times. The glitch and the fix were entirely too mysterious for cheers and champagne. "I've become sort of numb in the last few days," Laeser admitted.

Still the perplexing question of what had gone wrong in the first place remained. Laeser said he suspected that it was "crud in the gears," but added that the coincidence of the glitch and ring-plane crossing was highly suspicious. "I have trouble accepting that. I'm willing to bet that something else caused those missed pictures before ring-plane crossing."

While the controllers and anomaly teams wrestled with the scan platform problem, science continued as usual. In fact, the weird events at ring-plane crossing gave the scientists what amounted to an extra, unplanned experiment to ponder. Not all the strange data could be explained immediately, but a general set of ideas was beginning to emerge from the confusion.

Fred Scarf's plasma-wave experiment went crazy shortly before ring-plane crossing. The instrument, which normally monitors very-long-wavelength emissions, suddenly began registering waves on the order of a million times more intense than normal; the event came close to "saturating" the instrument. To Scarf, the tape of the event sounded "very much like things hitting the spacecraft. I think the most likely explanation is that these are impacts." Apparently tiny charged dust particles were annihilating themselves as they smashed into Voyager. All that was left of the dust after the collisions was a high-energy plasma which was seen in Scarf's strange data.

The planetary radio astronomy experiment also found "a very strong electromagnetic discharge," according to Joseph Romig. The result again suggested that the charged dust, traveling at 13 km per second, had smashed into the spacecraft, leading Romig to conclude: "Voyager 2 did not pass outside the rings of Saturn, but in fact passed *through* a ring."

That startling conclusion was the only plausible explanation for the ring-plane event, except for the fact that the PRA data extended some 750 km above and below the plane. Were there multiple, "stacked" rings, like phonograph records on a spindle? That seemed hard to accept; more likely, the PRA was registering some sort of wave phenomena which propagated vertically with respect to the ring plane.

If Voyager had, indeed, gone straight through a ring, which ring? The trajectory was dead on target, sending the spacecraft through what was supposed to be a broad gap between the G and E rings. There should not have been a ring anywhere near that location; on the other hand, rings were popping up in a lot of places where they ought not to have been.

Brad Smith reported the discovery of an exceedingly odd ring within the Encke Division. Voyager 1 had found a hint of clumpiness in the region; now, Voyager 2 images revealed a bright, narrow, and wildly eccentric ringlet which soon became known as the "Kinky Encke." Again, there was no sign of the embedded moonlets which had been presumed necessary to create such a structure within the rings. The only thing Voyager 2 saw in the rings was more rings; in the B ring, high-resolution images were showing thirty ringlets where Voyager 1 saw only one. The ring images had a resolution of less than 15 km; to see finer structure, one had only to take a glance at Lonnie Lane's half-mile-long strip chart.

When Lane came up to the Von Karman stage to announce his early results, his pride was evident. "It's taken me about three years to traverse the seventy-five feet from the back of the room to the front," he observed. The long struggle with the photopolarimeter had clearly been worth the effort. It was going to take "several years" to work through the mountain of ring occultation data, but it was already obvious that the number of ringlets was going to be enormous. Imaging had seen a few thousand ringlets; PPS may have seen more than a hundred thousand.

The rings were numerous, and they were extremely thin. Lane's

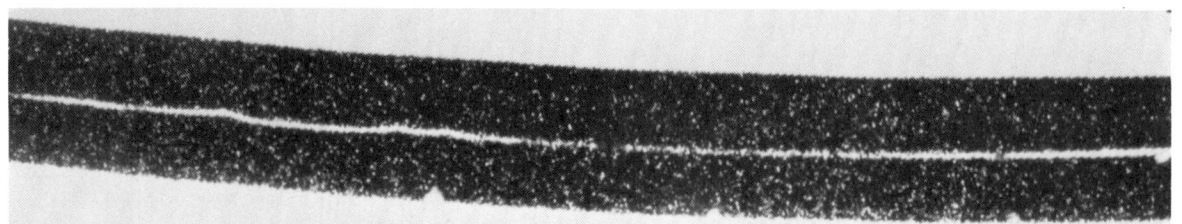

data for the A ring put an upper limit on ring thickness of just 150 meters. The actual thickness may have been as small as 100 meters.

The closer the scientists looked at the rings, the more detail they saw. In the F ring, for example, the PPS found as many as fourteen separate strands where imaging had seen just one. Scientists began to speculate that, at an extremely small scale, the rings might consist of individual particles lining up in Indian file—micromini-ringlets. The distribution of the ringlets was in some ways similar to the circular ripples generated by tossing a stone into a pond. Increasingly, the evidence was swinging toward wave phenomena as the explanation

The "Kinky Encke," a wildly eccentric ringlet within the Encke Division, was discovered in this image from 435,000 miles. The ringlet is about 9 miles wide.

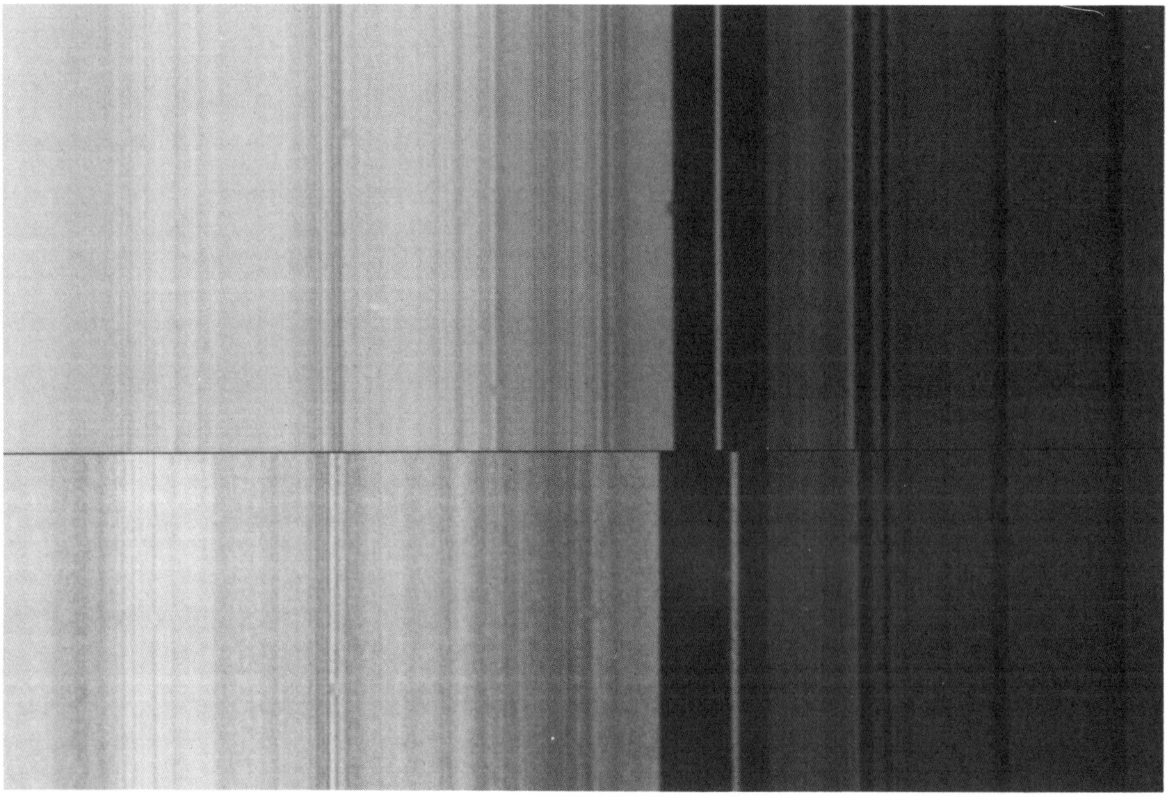

for the ring structure. Very long gravitational waves, generated by satellite resonances, could be dictating the spacing of the ringlets.

There were so incredibly many ringlets, and so much ring material even in the gaps, that the scientists were finding it difficult to live with the classical concept of discrete, individual rings. By the end of the encounter, an alternate explanation was being kicked around. Was it possible that the rings really consist of one flat, continuous sheet of particles? The waves rippling across the sheet would segregate particles into dense and less dense ribbons, but there would be virtually no place that was completely free of ring material.* If the continuous-particle sheet idea was accurate, then the scientists were going to have to rethink the entire concept of what a planetary rig is.

At Saturn, anything seemed possible. The braided F ring was a prominent example of the diversity on display at the sixth planet. Although some F-ring coverage was lost due to the stuck platform, two or three segments of the ring had been examined at high resolution. The imaging scientists had expected to find embedded moonlets. Instead, they found no moonlets and, rather shockingly, no braids. There was simply no sign of the complex structure seen by Voyager 1. Could the mystifying braids have been a temporary phenomenon? Possibly they were, but the idea of ephemeral structures in the rings was troubling. If the F-ring braids were transient, what else might be? Could the disputed sightings of the D ring be explained by supposing that it was a temporary, now-you-see-it-now-you-don't structure?†

And where did all this leave the spokes? The trend toward wave phenomena and electromagnetic interactions with fine particles was consistent with the best spoke theories, but there was much that still had to be explained. At ring-plane crossing, the PRA experiment had

---

* In this context, it was significant that in all the ring images there was no coverage of any individual ringlet all the way around the planet. We didn't *know* that each ringlet was, in fact, circular. To some observers, it seemed possible that the rings actually consisted of one very long, continuous spiral, like the groove in a phonograph record. It was, of course, an utterly preposterous idea—but then, this was Saturn.
† Time disposed of these and many other questions being asked during the encounter. Later analysis of some of the Voyager 2 images showed that the braids were still there, after all. They were not ephemeral, but they were still baffling.

Facing page: *Images from opposite sides of the planet are juxtaposed here, revealing the outer B ring and the inner edge of the Cassini Division. The mismatch between the two sides shows that the rings are not as uniformly concentric as had been supposed. One ringlet in the Cassini Division is offset by about 30 miles.*

observed an incredibly strong electrostatic discharge. It was similar to terrestrial lightning, but its source region was in the rings, at 1.81 R_s—in the heart of the B ring. The discharge was on the order of 1,000 megawatts, making it a thousand times stronger than similar emissions seen at Jupiter. The scientists couldn't say for sure if this powerful discharge or the others which had been observed were related to the spokes, but the circumstantial evidence pointing in that direction was now overwhelming.

All the various discharges and radio emissions coming from Saturn were puzzling, one way or another; tying them all together in one neat, unified theory was clearly going to be a challenge. The task didn't get any easier when, shortly after the spacecraft passed the planet, the periodic Saturnian radio pulse disappeared. "For some reason," said Fred Scarf, "Saturn turned off for four days." "This dead silence," added another scientist, "is just incredible."

The radio emission seemed to be tied to the rotation of Saturn's magnetic field. Its mysterious shutdown implied an irregularity in the field. Norm Ness had been looking for—and more or less expecting —some anomaly in the shape or strength of the field; he found none. The 0.7° tilt of the field was the only hint of irregularity, but, said Ness, "One is hard-pressed to understand how such a small tilt could have *any* effect in the modulation of the PRA signals." The emissions, like their sudden shutdown, remained a mystery.

Many such mysteries might come unraveled if we knew more about Saturn's interior. Detailed imaging of the cloud tops had provided a mass of new data about convective regions, ovals, and jet streams; there was even enough symmetry visible to give new life to Ingersoll's cylinders. But what was happening beneath the cloud tops?

At the final briefing on August 30, Ingersoll explained that any model of Saturn, his own included, would have to "agree with all of our observations." However, there was no computer big enough to simulate all the complex activity that had been observed, "so we have to make approximations." It would be necessary to approximate the structure of the atmosphere and the interior, and then "make a clever assumption about the interface between the two." Such a clever assumption, ideally, would account for the zonal jets, the atmospheric temperature gradients, the flow patterns around the ovals, the existence of long-lived ovals, and the complicated turbulence in the eddies and streams. In addition, the model ought to work for both Jupiter and Saturn. "I'm very optimistic," Ingersoll concluded, "that we have enough data, especially with Voyager 2, so that most models are going to fail to explain one or more of these observations." Ingersoll added that he expected that theories would be falling like tenpins for years to come.

Saturn's interior would probably not be well understood until a Galileo-style mission could provide us with information about what was going on beneath the cloud tops. "We have a lot of interesting problems that we can't really get at with a flyby mission," said Ed Stone. "One very much would like to have a sample of Saturn's atmosphere, and that means an entry probe." Such a mission had already been planned for the eighties and nineties; now it seemed doubtful that it would ever be flown.

Another highly desirable mission would be a Titan lander, but funding for such a mission would probably be even less likely than for a Saturn entry probe. Voyager 2 did not come as close to Titan as its sister ship, but it did confirm some previous observations and add one more unexpected scrap of knowledge. Titan, said Len Tyler, apparently had no ionosphere. "This is the first case where we have found a body with an atmosphere without an ionosphere. This is a mystery."

Another Titan mystery, though, seemed to be on the verge of solution. Ever since Kuiper identified methane in Titan's spectrum in 1944, scientists had speculated about the possibility of complex organic compounds forming in the Titanian atmosphere. Voyager 1 had identified at least ten hydrocarbon compounds in Titan's dense clouds, ranging from ethylene to methyl acetylene. It was a fascinating mix of organics, but it may only have been the tip of the iceberg.

At a background session on the 27th, Carl Sagan stepped back into the scientific limelight. After the success of *Cosmos*, some observers doubted that Sagan would ever be able to get back to workaday science. But Sagan was more than a little weary of the demands of celebrityhood; looking much more relaxed than he did in the days of *Cosmos* and obviously enjoying himself enormously, Sagan reported the results of a fascinating experiment.

Working with B. N. Khare at Cornell, Sagan attempted to

duplicate the Titanian atmosphere in an experiment that was similar to the famous Miller–Urey "primordial soup" experiments in the fifties. Miller and Urey had concocted a "model" of Earth's early atmosphere and exposed it to an electric spark for a period of weeks to months. The products of the brew included a variety of hydrocarbon compounds considered necessary for the creation of life. Sagan and Khare duplicated the experiment using a model of Titan's nitrogen–methane atmosphere based on early Voyager 1 results. The gases were bottled up and exposed to an electric spark for a million seconds, or about four months.

Titan's thick atmosphere bends light around to the night side, creating a crescent moon effect. Titan's atmosphere is denser than Earth's and is composed mainly of nitrogen.

Before long, a thin red film appeared on the walls of the test tube. The film was composed of organic matter synthesized from the atmospheric gases. "It's a very complicated kind of material," said Sagan, very long, nonrepeating organic molecules known as heteropolymers. The material was the same color as the clouds of Titan and matched almost precisely the polarization and reflection properties of the Titan smog layer seen by Voyager 1. "It's just bang on," said Sagan. "No other material we've made over the years has that kind of match." His conclusion: "There's some chance that we've bottled the clouds of Titan."

Sagan stressed that the results were still very preliminary. There remained the arduous task of breaking down and analyzing the precise components of the heteropolymers, which Sagan called "tholins," a coined word derived from the Greek and meaning (said Sagan), "star tar." The tholins were certainly not of biological origin, but they could conceivably be the stuff of prebiology. If the raw materials of life were being mass-produced in a hostile environment like Titan, perhaps it was an indication that life was abundant throughout the universe. Implications aside, the experiment provided scientists with an exciting new perspective about conditions on Titan.

Two-toned Iapetus was another moon concerning which Sagan seemed to be on the right track. "There is a sense of ongoing debate about what these pictures mean," said Sagan, but it was beginning to look as if the dark material on Iapetus consisted of "complex organic goo." As experimenter Von Eshleman put it, the dark material on Iapetus may look as dark as pitch because it may actually *be* pitch. The combination of ice and asphalt gave rise to a picture of Iapetus as the ultimate ski resort: icy slopes on one half of the planet and a paved parking lot on the other half.

The original source of Iapetus' dark organics may have been methane (CH_4) from the solar nebula or carbonaceous chondrites. However, it was still unclear whether the dark material came from the interior or from Phoebe or from some as yet unidentified process. "The debate, I'm sure, will continue to rage," said Sagan. The sharp boundary between terrains seemed to imply internal processing, yet the dark hemisphere was almost precisely centered around the leading edge of the moon, a circumstance which seemed to demand external origin. How could a purely internal process "know" where the center of the leading hemisphere was?

Long-range imaging of Phoebe did little to resolve the controversy. With its scan platform still behaving erratically, Voyager 2 managed to send back a few final postcards from Saturn as it headed into interplanetary space. The Phoebe images showed a small (200 km) and surprisingly round moon (for such a small body) with a dark surface and some indications of cratering. If Phoebe was sputtering dark dust into space, there was no obvious evidence of it.

Saturn's other moons would also remain a subject for debate and head scratching. Parts of Enceladus's crazy-quilt surface seemed to be amazingly young. "Our guess," said Larry Soderblom, "is that this patch of terrain had to be formed 100 million years ago or less.... It's very unlikely for a planet to run geologically for ninety-eight percent of its history, and then suddenly stop." The inescapable conclusion was that Enceladus is still geologically active. Some sort of Io-style tidal heating by Saturn and the other moons may be the reason for the satellite's apparent activity.

Tethys, on the other hand, seems to be quite dead. The huge trench is probably the result of the expansion of ice. In its early history, Tethys probably froze from the outside in, and in the process the surface area expanded by about five to ten percent. Since the trench comprises about ten percent of the moon's total surface area, it seems likely that the planet simply cracked as the ice expanded.

At the final press briefing, Larry Soderblom recalled that just three years earlier, in 1978, we had seen the surfaces of only five planetary bodies. "We thought we understood a little about how planets behave." Now, the number of planetary surfaces seen by terrestrial eyes numbered twenty-four, and scientists were coming to realize that their earlier ideas about the behavior of planets were simply inadequate. Who could have predicted the bizarre surface of Io, Ganymede, or Enceladus? And what would the moons of Uranus and Neptune look like? The possibilities seemed endless. Said Soderblom, "There is an infinite variety yet to be explored."

Soderblom's words applied not just to moons, but to the entire range of planetary phenomena, from mountains to magnetospheres, from red spots to rings. If they accomplished nothing else, Voyagers 1 and 2 had conclusively demonstrated how limited our earthly perspective really is. The Voyagers had given us answers, but more, they had given us new questions to ask. As Andy Ingersoll put it, "The world will be a very dull and somewhat grim place if we stop posing questions and discovering new questions we can't even pose."

But "dull and somewhat grim" turned out to be an accurate description of the proposed NASA budget for fiscal year 1983. After months of speculation and apprehension, the Reagan Administration's budget recommendations were released in February, 1982. NASA had requested $7.6 billion, but it would receive just $6.6 billion, most of it earmarked for the shuttle. The planetary exploration budget was slashed from an already minimal $205 million to a bare-bones figure of $154.6 million.

The Galileo mission to Jupiter was saved, although launch was delayed until 1985. Uncertainties about the launch vehicle forced a redesign of the mission, and the probe would not reach Jupiter until 1990. That was the good news.

All else was desolation. There would be no American mission to Halley's Comet. The Venus orbiting imaging radar mission was canceled. All the surviving Pioneer spacecraft, including Pioneers 10 and 11, bound for the edge of the solar system, would be turned off. Funds for the analysis of data already collected were cut by nearly twenty-five percent. And there would be no new missions to the planets.

"There is no intent to stop the planetary work," insisted Administrator Beggs. Some observers gave Beggs high marks for putting up an unexpectedly good fight on the budget; there were even a few minor increases in some areas, such as physics and the life sciences. The massive hemorrhages some had feared were prevented, it seemed; but a hundred small cuts can be as deadly in the long run as a severed artery.

Beggs said that space exploration had received "an enormous amount of both public and private support . . . a very heartwarming and gratifying response." But the public had much more on its mind in the troubled winter of '82 than space exploration. The nation was in a deep recession, unemployment was soaring, and the President was sticking to his guns in his quixotic quest to make Reaganomics work. In the face of a projected $100 billion deficit, Congress was showing signs of rebellion. Further budget cuts seemed all but inevitable. "Given the tightly constrained fiscal environment in which this budget was prepared," said Beggs, "I believe we did well." But a budget request is not the same thing as a budget, and NASA seemed especially vulnerable. Few in government seemed willing to question the bloated defense budget, yet funds for essential social programs had already been reduced beyond what many considered to be an acceptable minimum. If it came down to a choice between NASA and hot lunches for impoverished children, there could be only one humane decision.

To pose the problem in those terms, however, was misleading. The paltry sums slated for NASA would make little difference to the health and well-being of the millions of Americans who were struggling to survive the hard times of the eighties. The real choice was between programs designed to preserve life and programs designed to end it.

Our small, blue planet is armed to the teeth. The greatest nation on that planet seems to have come to the decision that it cannot afford to house, feed, and educate its population, but that it *can* afford to expand and export the most awesome arsenal ever assembled. As the Voyagers sail onward toward cold, barren, alien worlds, they leave behind them a warm and bountiful world that is poised on the brink of catastrophe.

The Voyager 2 Uranus encounter was preserved in the proposed NASA budget. The spacecraft was ailing (the scan platform gearbox seemed to be approaching the end of its operational lifetime, the imaging system was slowly degrading, and the onboard computer memory seemed to be suffering from a sort of progressive amnesia), but the Voyager team was still confident that they could keep their vehicle alive until 1986. Our questions about Uranus would be answered; but questions about the Earth remained. Unless we could find a way to live with ourselves and reverse our deadly march toward nuclear confrontation, January, 1986, might find Voyager 2 beaming back data from one lifeless world to another.

Even if we avoid that ultimate mistake, it seems clear that the Golden Age of Solar System Exploration has come to an end. The last Golden Age of Exploration, at the end of the fifteenth century

An artistic montage of the Saturnian family: Saturn rises behind Dione, with Tethys and Mimas to the right, Enceladus and Rhea to the left, and Titan alone at upper right. Despite the many tantalizing mysteries revealed by the Voyagers, there are no plans to return to Saturn in this century.

and the beginning of the sixteenth, was also short-lived. The voyages of Columbus and Magellan soon gave way to the monstrous conquests of Cortez and Pizarro.

Like the Spanish, Portuguese, and English before us, we seem intent on exploiting our new discoveries before we fully understand them. "Looking to the future," said Beggs in his budget message, "I am confident we will make the most of the opportunities the shuttle affords as we open a magnificent new era of transportation, commerce and industrialization in space."

But there is reason to hope that as we exploit this new wilder-

ness, we may avoid the tragic mistakes of the past. Perhaps five centuries of bloody competition for new land and resources have taught us something. Certainly, we have learned something new and important from the heroic odyssey of the Voyagers.

"I believe," said Larry Soderblom following the Voyager 2 Saturn encounter, "that man's concept of the universe has been irreversibly changed." One had a sense of a great door slowly swinging open to reveal a garden of unearthly delights stretching all the way to the farthest horizon. The last time mankind experienced such a sensation was five hundred years ago, when the beauty, wealth, and diversity of Planet Earth were first unveiled. Those discoveries helped propel civilization into an unprecedented era of intellectual and cultural enlightenment. Perhaps it is not too much to hope that the Voyagers may do the same for our own troubled age.

The wonders and wealth of the Americas, like the glories of Jupiter and Saturn, were important in and of themselves. But even more important is the fact that human beings are able to discover those wonders, to see with their own eyes vistas never before imagined. "In a sense," said Soderblom, "we *are* Voyager; in that sense, human beings now measure a billion kilometers in dimension."

With the perspective of billion-kilometer beings, we can see much that was hidden from us before. We can perceive, for the first time, our special responsibility to ourselves, our planet, and to the universe itself. Other, wiser beings may be waiting for us in the stars, but, for now, we are alone in Wonderland, the sole custodians of an ancient and awesome solar system. Voyager gave us a glimpse of all that lies beyond us, and the *experience* of Voyager gave us a new appreciation of what is within us. We are, in a dark and challenging universe, a bright spark of intelligence, a solitary brilliance glowing with hope and compassion, curiosity and courage—as bright and glittering and promising as Jupiter and Saturn themselves, shining and sparkling and calling to us from the endless sky.

INDEX

Underscored numbers refer to original art.

albedo (*see also* names of planets, moons), 96
Alexander, George, 121
Amalthea, <u>40</u>, 41, 101–2, 103, 112, 138
   interaction with Jupiter, 147
   Io, ejecta from, 161
   location, 93–94
   surface, 161
American Astronomical Society, 145, 181–85, 222–23
Ames Research Center, 165 ff.
Antoniadi, E. M., 199
Apollo, missions and craft, 5, 6–7, 9, 13
asteroid
   belt, crossing of, 173
   evolution of satellites and, 232–33
atmospheres (*see also* specific planets), 23, 24
   circulation, 25
   instrumentation and study of, 53–54, 55, 60
Armstrong, Neil, 7

Barnard, E. E., 41
Baum, William, 186
Beatles, 76
Beatty, J. Kelly, 121, 199–200, 232
Beebe, Rita, 141, 212, 233, 234
Beggs, James Montgomery, 220–24 *passim*, 242, 259–60
Bergman, Jules, 126–27
Berry, Richard, 121
Bessel, Freidrich Wilhelm, 19
Bradbury, Ray, 102
Bridge, Herbert S., 58
Bristow, Frank, 120, 121
Broadfoot, Lyle, 54, 93, 113
Brown, Jerry, 103
Brown, Robert, 76

263

INDEX

Bunker, Anne, 193
Burke, B. F., 27

Callisto, 40, 41, 42, 93, 96, 138, 151, 196
  albedo, 96
  approaches to, closest, 111, 135
  cratering, 108, 111, 138, 153–54
  Ganymede compared with, 154–56
  orbit, 66
  structure, 154
  surface, 95–97, 100, 111, 153 ff.
  water, 154
canals
  Europa, 136–37
  Mars, 11, 12, 136–37
Carter, Jimmy, 77
Casani, John, 74, 133
Cassen, Pat, 116, 159
Cassini Division, *see* Rings, Saturn
Cassini, Giovanni Domenico, 17, 18–19, 33, 42, 43
Chiron, 232
Clarke, Arthur C., 44, 82, 102
comets, 182, 221, 232–33, 259
communication, interstellar, 69–82, 114
communications systems, spacecraft, 62–63, 172, 245
  modification of, 85
  equipment problems, 85, 131, 133, 138, 141, 167, 168
  interference, 176–77
  time lags, 103, 172
  weather problems, 99, 167, 168, 176–78
computer systems
  Mariner, 62
  Pioneer, 87, 167, 172
  timing, 103, 107, 108
  Viking, 62
  Voyager, 62–63, 99, 107, 108, 167, 193, 248
Congress, 77, 182, 222, 223, 260
Cook, Allan, 199, 213
Cooke, Bob, 121
Copernicus, 40

cosmic rays, 59
Cosmos (Sagan TV series), 127–29, 187
cratering (*see also* names of planets, moons), 21, 95–96, 111, 153, 213, 232
Crespi, Donati, 19
Crippen, Robert, 7
Cruikshank, Dale, 185
Cuzzi, Jeff, 216

Davis, Esker (EK), 85, 141, 246, 248–49
Deep Space Network, 62–63, 222
  radio uplink, 85
  signal reception problems, 99, 168, 176–78
Defense Department, 7
Dickinson, Angie, 135
Dione
  co-orbital moon, 115–16
  cratering, 200
  Enceladus affected by, 213
  and ring formation, 36–37
  evolution, 208
  surface, 201, 207
Division for Planetary Science of American Astronomical Society, 145, 181–85, 222–23
Drake, Frank, 71–72, 74, 75, 79
Druyan, Ann, 74, 75, 78
Dyson, Freeman, 72

Earth, 5, 10
  cratering, 95, 214
  formation, 21
  magnetic fields, 27, 28
  magnetosphere, 57–58, 167
  moon, *see* Moon, Earth's
  observations, astronomical from (*see also* names of planets, subjects), 10–12, 17–20, 26–27, 219
  oceans, atmospheric analogy to, 33
  organic compounds and/or organisms, 31
  radio emissions, 210
  solar wind, 57–58

INDEX

tectonics, 156
volcanism, 116, 158, 159
Eberhart, Jonathan, 103, 120–21, 123
electromagnetism (*see also* subjects)
   instrumentation for study of, 53–55, 56
Enceladus, 208, 213, 242, 247
   age, 243, 258
   cratering, 213, 232, 243
   rings, Saturn's, and, 36–37, 186
   Saturn, effect of, 214
   terrain, 213, 243, 258
Encke Division, 251
Eshleman, Von R., 60, 208, 209, 257
Esposito, Larry, 237
Europa, 40, 41–42, 96, 100, 116, 138, 151, 156
   albedo, 96
   approach, closest, 135
   canals, 136–37
   cratering, 137, 157
   ice crust, thickness of, 156
   Io, effect of, 159
   life on, 157
   ridges, 157–58
   structure, 135–36, 138, 157
   surface, 95–96, 136–38, 142–43, 157–58
   water, 156, 157
European Space Agency, 221
Explorer I, 27

Ferris, Timothy, 74, 75, 76
Franklin, F. L., 27
French Division, 164
Frosch, Robert, 140, 182

galaxies, *see* Stars and star systems
Galilean satellites (*see also* Jupiter, moons; names), 41–44, 65–66, 81, 93 ff., 144, 151 ff., 189
Galileo, 17, 18, 34, 35, 39–41, 219
Galileo mission, 150–51, 182, 221, 259
Ganymede, 40, 41, 42, 93, 96, 100, 151
   albedo, 96
   approaches to, closest, 107, 108, 135

Callisto compared with, 154–56
   cratering, 108–9
   mountains, 156
   Rhea and Dione, compared with, 201
   ridge systems, 110
   size, 210
   surface, 95–97, 108–11, 112, 135, 138–39, 154–56
   tectonics, 110, 156
   temperature, 156
Gehrels, Tom, 167, 169, 174, 179
Guerin Division, 164, 179
Guerin, Pierre, 164
Gurnett, Dale, 210

Hall, Asaph, 37
Hall, Charles, 172, 173
Halley's Comet, 182, 221, 259
Hanel, Rudolph, 53, 113, 208–9
Hanson, Richard S., 32
Hartmann, William K., 43–44
Heacock, Ray, 132–33, 138, 203, 246
Heinlein, Robert A., 82
Herschel, Sir John, 42
Herschel, William, 35, 40
Hibbs, Al, 90, 102–3, 104, 106, 231, 245
Hinners, Noel, 45–47, 107
Hooke, Robert, 18, 33
Horowitz, Norman, 124
Hunt, Garry, 97, 132, 233
Huygens, Christian, 34–35, 42, 219
Hyperion, 239–40

Iapetus, 43–44, 129, 213, 248
   approach, closest, 229–31
   cratering, 231
   mass, 242
   organic compounds, 230, 242, 257–58
   Phoebe, dust from, 202, 229, 231, 257–58
   Pioneer 11 mission, 168
   reflectivity of hemispheres, 43–44, 202, 213–15, 229–30, 231
   trajectory, Voyager, 229, 230

265

INDEX

imaging systems
  Mariner, 50
  Pioneer, 86, 87, 89, 165, 166–67, 176
  Viking, 50–51
  Voyager, 51–53, 62, 88 ff., 101, 104, 107, 166, 193, 225–26, 246–50
infrared spectroscopy, 53, 56, 113, 149, 175 ff.
Ingersoll, Andrew, 233–35, 254, 259
  Jupiter, 32, 38, 97, 114, 126, 141–43, 148, 149
  Saturn system, 175, 177–78, 193–94, 195, 209, 211, 212, 233–35, 254–55
instrumentation, scientific, Pioneer, 165–66, 167, 172–73
instrumentation, scientific, Voyager (*see also* subjects), 49–68, 114, 165, 173, 224–26
  problems and failures, 84–86, 107, 108, 131–32, 133, 138, 141, 224–25, 246–50, 258
  radiation effects, 66, 86, 87, 92, 98, 107, 131–32, 138
Internation Astronomical Union (IAU), 175
International Solar Polar Mission, 221
Io, 40, 41–42, 93, 94, 96, 100, 102, 103, 111–12, 138, 151, 158, 191
  albedo, 96
  approaches to, closest, 98 ff., 104 ff., 134
  craters, 100, 101, 104–9 *passim*
  flux tube, 57, 107–8
  interactions with Jupiter, 56–57, 93, 107, 145, 147, 161, 212
  mountains, 105, 160
  orbit, 66, 159
  surface, 94 ff., 105–6, 107, 129, 160–61
  temperatures, 158–59
  torus, 57, 92–93, 134, 147
  volcanism, 106, 115–16, 134, 137, 142, 145, 158–61
  water, 106

Janus, 174, 185
Jeffreys, Harold, 25
Jet Propulsion Laboratory (*see also* names of persons; subjects), 182
  data processing, 63
  press briefings (*see also* Media, coverage by), 90, 119 ff., 186 ff., 220 ff., 242
  imaging system, 50–53, 80
Johnson, Torrence, 136, 139, 159, 161, 190, 196, 197, 231
Jupiter, 14, 16–19, 112
  analyses of system, 146–61
  approaches to, closest, 98 ff., 103–4, 133, 139 ff.
  atmosphere, 22–33 *passim*, 37–39, 87, 88 ff., 97–98, 104, 133, 141–42, 147–51, 187
  auroras, 113, 147
  bowshock, 91–92, 134
  communication time to, 62
  composition, 22–25, 30–31, 149
  electromagnetism, 27–30, 42, 56, 57–59
  formation and evolution, 22, 23, 29, 39, 41–42, 151–53
  Galileo mission, 150–51, 221, 259
  gravity assists at, 63, 115, 163
  Great Red Spot, 11, 18–19, 30, 88–89, 90, 94, 97, 104, 113, 133, 149–50, 234
  interior, 23, 24, 148
  magnetic fields, 26–30, 87
  magnetosphere, 27, 28, 56–57, 91–93, 133–34, 145–47, 161, 167, 212–13
  mass, 16, 18, 19
  moons (*see also* names), 22, 39–42, 66, 87, 92 ff., 134 ff., 146–47, 188
  night-side view, 142–43
  observations from Earth of, 11–12, 17–20, 30–31, 114, 219
  organic compounds and/or organisms, 31–32
  Pioneer mission, 70–71, 86–87, 89, 91, 163, 166, 171
  radiation, and effects on spacecraft,

INDEX

66, 86, 87, 92, 98, 107, 131–32, 134, 138
radio emissions, 27–28, 42, 98, 113
ring, 101, 113–14, 143–44, 161, 166, 168, 174
rotation, 17–18
Saturn compared with, 37–39, 187 ff., 211 ff.
solar wind, 91, 92, 147
sounds of, 98
symmetry, hemispheric, 233–34
symposium on, 102
temperatures, 25, 28–30, 39, 41–42, 87, 96, 113, 146–47, 149, 212
trajectories to, 63–67, 84–85, 101
vacuum, magnetosphere, 147
Voyager 1 mission, 14, 48 ff., 83 ff., 127, 129–30, 133–34, 138, 166, 173, 249
Voyager 2 mission, 14, 30, 48 ff., 83–85, 91, 131 ff., 165, 173, 224–25

Kennedy, John F., 6, 45
Khare, B. N., 255–57
Kieffer, Sue, 158
Kohlhase, Charles, 64–65, 67, 84, 103, 202, 245, 247
Kowal, Charles, 232
Krimigis, Stamatios (Tom), 59
Kuiper, Gerard, 42, 195, 255

Laeser, Dick, 246, 247, 249–50
Lane, Arthur L. (Lonnie), 225, 245, 251
Lenorovitz, Jeff, 103
Libby, Willard F., 32
life and/or organic compounds, extra-terrestrial, 32, 118
  communication attempts, 69–82, 114
  instrumentation for detection of, 53
  Iapetus, 230, 242, 257–58
  Jupiter, 31–32
  Mars, 11, 12, 13, 32
  Titan, 43, 195, 204, 209, 255–57

Lomberg, Jon, 74, 75, 79, 80, 81, 82, 93
Loudon, Jim, 110–11
Low, Frank, 28
Lowell, Percival, 11, 136–37

MacElroy, R. D., 32
magnetopause, 91
magnetospheres (*see also* names of bodies), 28, 171
  bowshock, 91–92
  instrumentation for study of, 57–59
  solar wind and, 57–58, 91, 147, 170–71
Mariner missions and spacecraft, 47, 60, 61, 184
  Mars, 12, 13–14, 50, 62, 184
  Mercury "moon," 125
  press coverage, 125
  symposium, 102
  Venus, 12
Marius, Simon, 40–41
Mars
  atmosphere, 11
  canals, 11, 12, 136–37
  cratering, 12, 95, 153
  landings on, 13–14, 191
  life on, 11, 12, 13, 32
  Mariner missions, 12, 13–14, 50, 62, 184
  moons, 37, 39, 95
  Mutch memorial, 140
  observation from Earth of, 10–11
  polar caps, 11
  Soviet exploration of, 12
  terrain, 13–14, 129
  Viking missions, 13–14, 32, 50–51, 191
  Voyager mission, 48
Mars I, 12
Masursky, Hal, 96, 106, 135, 136, 137, 138, 229–30, 231, 243
Maxwell, James Clark, 36
McKinley, Ed, 64
media, coverage by
  Pioneer missions, 170 ff.
  Voyager missions, 90, 99 ff., 118–30, 132 ff., 144, 186 ff., 203 ff.,

267

INDEX

media, coverage by, Voyager
    missions (*cont.*)
        208, 210, 219, 220 ff., 231, 238,
        242 ff., 254–55, 258
Menzel, Donald, 25
Mercury, 39
    atmosphere, 10
    cratering, 21, 95, 153, 213
    observation from Earth of, 10
    Mariner mission and "moon" of, 125
meteorites, 21, 230
meteoroids, detection of impacts with,
    172–73
Michener, James, 183–84, 224
Miller, Stanley, 31
Mimas
    cratering, 207, 213, 233, 242
    and ring formation, 36–37
moon, Earth's, 40, 95, 96
    albedo, 96
    Apollo missions, 6–9 *passim*, 13
    cratering, 21, 95, 153
moons (*see also* names of planets,
    moons), 39
    formation and evolution, 39, 196,
        201, 213, 231 ff., 242
    Galilean satellites, *see* Galilean
        satellites; Jupiter, moons; names
    rings and, 36–37, 39, 189–90,
        197, 198, 217–18, 235–37
Morabito, Linda, 115
Moral Majority, 224
Morrison, David, 14–15, 222
    Neptune system, 93, 97, 101,
        108–9, 126, 158, 172
    Saturn system, 190, 196, 198, 231
Murray, Bruce, 102, 129, 183
Mutch, Thomas A. (Tim), 140–41,
    181

NASA (*see also* names of persons,
    programs; subjects), 181 ff.,
    220–24
    budgets and funding, 7, 46–47,
        48–49, 129, 151, 182–84, 187,
        221–24, 259–60
    communication, interstellar, 70 ff.
    goals and motivations, 46 ff.

    radio frequencies, 176–77
    Skylab, 132
National Enquirer, 118
Neal, Roy, 126
Ness, Norm, 207, 209, 254
Neptune, 39, 113
    observations from Earth of, 11–12
    travel time to, 62
    Voyager mission, 14, 63, 65, 67,
        246–47
Ness, Norman F., 58
Newton, Sir Isaac, 80
Nixon, Richard M., 6
nomenclature, 40–41, 42, 174–75,
    185*n*
Norcia, Anne, 103

occultation, 60
O'Meara, Stephen, 199–200
optical navigation, 67, 115, 245
organisms, *see* Life and/or organic
    compounds, extraterrestrial
Orlando, Tony, 135
O'Toole, Tom, 121, 132
Outer Planets Grand Tour, 48–49, 63
Overbye, Dennis, 121, 123
Owen, Toby, 211

Parks, Robert, 107, 132–33
particle beam astronomy, 173, 178
particles and fields studies (*see also*
    subjects)
    Pioneer, 165–66, 171 ff.
    Voyager, 46–49, 93, 107–8, 113,
        145–47, 166
Peale, Stanton, 116, 159
Pearl, Jack, 116
Peck, Gregory, 135
Perlman, Dave, 121
Phoebe, 44, 248, 249, 258
    cratering, 258
    Iapetus, effect on, 202, 229, 231,
        257–58
photopolarimetry, 55, 56, 224–25, 245
    failure, 85
Pieri, David, 161
"Pioneer Gap" or "Division," 174–75,
    179

268

INDEX

Pioneer missions and spacecraft (*see also* names of planets, moons; subjects), 14–15, 47, 86, 163 ff., 259
 communication, interstellar, 70–74 *passim*, 114
 end of, 179–80
 goals and instrumentation, 165–66, 167, 172–73
 media briefings and coverage, 170 ff.
 problems and failures, 167
 trajectories, 163–65, 168
"Pioneer Rock," 178, 185
Planetary Society, 129
planets (*see also* Solar system; names, subjects), 111
 alignment, 48
 distances to of spacecraft, 87n
 formation and evolution, 20–22, 213
 twin, 154
plasma experiments, 58–59
Pluto, 12, 14, 39, 48
Pollack, James, 43, 199

radio
 communications with spacecraft, 85, 131, 132, 138, 141, 176–77
 emissions, planetary, 27–28, 42, 166, 186, 200, 210, 218, 238, 254
radio astronomy, 26–30, 42, 54–55, 60, 186, 195, 250–51
Ramsey, W. H., 25
Ranger mission, 47
Reagan, Ronald, and Administration, 183, 186–87, 221–24, 259–60
Reynolds, Ray, 116, 159
Rhea, 200–1
ring(s), Jupiter, 101, 113–14, 143–44, 161, 166, 168, 174
ring(s), Saturn, 12, 113, 114, 178–79, 185, 189, 197, 215, 216, 225, 234
 braids, 204–5, 253
 Cassini Division, <u>34</u>, 35–37, 169, 174, 189, 197–98, 216, 217, 236, 237

 emissions, 218, 254
 Encke Division, 252
 formation and evolution, 39, 169, 189–90, 217–18, 232, 235–39, 253
 Guerin (French) Division, 164, 179
 identification, 35, 178
 moonlets/satellites, 186, 217–18, 227, 235–37, 252, 253
 observations from Earth, 34–37, 164, 170, 179, 186, 199–200, 215–16
 particles, 169–70, 173–74, 198, 216–17, 237, 253
 "Pioneer Gap" or "Division," 174–75, 179
 ringlets, 189, 217, 252–53
 ring-plane crossings, 163–65, 171 ff., 229, 245, 248, 250–54
 spokes, 190–91, 198–200, 218, 228–29, 238, 253–54
 temperatures, 175
ring(s), Uranus, 113, 114, 174
Romig, Joseph, 250–51
Russia, *see* Soviet Union

Sagan, Carl, 30–31, 102, 127–30, 135, 204, 219, 223, 230, 242
 communication, interstellar, 71, 73–75
 Titan atmosphere experiment, 255–57
Sagan, Linda Salzman, 74, 75
Sagan, Nicholas, 77
Salisbury, Dave, 121
Saltus, Richard, 121
satellites, natural, *see* Galilean satellites; Moons; names
Saturn, 33–34
 "Anne's Spot," 193, 227, 235
 approaches, closest, 170–75, 186 ff.
 atmosphere, 22, 37–39, 166, 168, 174, 175, 187, 193–94, 211–12, 226–27, 234–35
 auroras, 175
 bowshock, 167, 170–71, 207, 239
 communication time to, 62
 composition, 33, 37–39

269

INDEX

Saturn (*cont.*)
    formation and evolution, 21–22, 24, 39, 175, 196
    gravity assists at, 63, 172, 229
    hemispheres, symmetry, 233–35
    interior, 38–39, 254–55
    Jupiter compared with, 37–39, 187 ff., 221 ff.
    magnetic field, 171, 200, 238, 254
    magnetosphere, 166, 167–68, 170–71, 200, 207, 209, 212–13, 239
    mass, 16, <u>18</u>
    moons (*see also* names), 22, 36–37, 39, 42–44, 168, 174, 178, 185–86, 188, 194 ff., 200 ff., 213, 215, 217–18, 227 ff., 235 ff., 251
    observations from Earth (*see also* Rings, Saturn), 11–12, 33–39 *passim*, 174, 185–86, 219
    Pioneer mission, 14, 44, 70–71, 132, 163 ff., 207, 215
    "Pioneer Rock," 178, 185
    polar region, north, 174
    radiation, 166
    radio emissions, 55, 166, 186, 200, 218, 238, 254
    rings (*see also* Rings, Saturn), 12, 34 ff., 113, 114, 163 ff., 178–79, 185, 186, 189–91, 197–200, 204–5, 215–18, 225 ff., 232 ff., 250 ff.
    rotation, 171, 186
    seasons, 212
    "shepherding satellites," 186, 227–28, 237
    solar wind, 167, 170–71
    temperatures, 37, 39, 175, 212
    trajectories to, 63, <u>67</u>, 84, 115, 133, 141, 168, 245
    Uranus trajectory and, 63, 164
    Voyager 1 mission, 35, 48 ff., 60, 163–64, 186 ff., 226–27, 238
    Voyager 2 mission, 48 ff., 60, 133, 163–64, 200, 220, 224 ff.
    scan platform, 60–61
        problems, 84, 85–86, 246–50, 258
Scarf, Frederick L., 59, 98, 146, 200, 238, 250, 254
Schweikart, Rusty, 103
Science, 116, 159
Shemansky, Donald, 92–93
Shepard, Alan, 7
Shklovsky, S. I., 27
Shoemaker, Gene, 231–33, 242
Simpson, John, 151, 173–74
Skylab, 132, 135
Slipher, V. M., 23
Smith, Bradford A., 52, 119, 126, 179, 251
    Jupiter system, 89, 94, 107, 112–16 *passim*, 126, 133, 135, 142, 151
    Saturn system, 126, 185–201 *passim*, 204–5, 208, 215–16, 227, 228, 231, 235–40 *passim*, 245, 251
Soderblom, Laurence, 119–20, 126, 201, 213, 218, 258
    Jupiter mission, 94–95, 99, 101, 104–6, 111–12, 135–36, 137, 138, 144, 201
    Saturn system, 201, 203, 205–8, 214, 232, 258, 262
Soffen, Gerald, 14
Space Age, overview of, 3–15
space exploration and programs (*see also* names, subjects), 140, 141, 181–85, 220–24, 258–62
    evaluation and interpretation of data, 144–45
    goals and motivations, 6–10, 45 ff., 70, 129–30
    instrumentation for, *see* Instrumentation; subjects
    landing on planet, first, 12–13
    moon landings, manned, 6–7, 8, 13
    overview of, 3–15
    and public attitudes, 5–10, 70, 118 ff., 127–30, 183–84, 187, 220 ff., 259–60
Space, interstellar
    cosmic rays, 59
    spacecraft in, 67–68, 71, 81–82
Space Shuttle, 7, 151, 182, 187, 259
Sputniks, 11
stars and star systems (*see also* Sun; names, subjects), 69

INDEX

double and multiple, 22
life, possibility of, and
    communication, 69–82, 114
and optical navigation, 67, 115, 245
spiral galaxies, 237
Stockman, David, 222
Stofan, Andrew, 221
Stone, Edward C., 50, 114, 126, 138,
    186, 187, 188, 203, 221, 224,
    247–48, 255
Strobel, Darrel, 209
Sullivan, Walter, 102
solar system (*see also* planets, subjects),
    10–12, 16, 111, 114
    composition, 22
    cosmic rays, 59
    formation and evolution of, 20–22,
        24, 39, 153, 218, 231 ff.
    nomenclature, 40–41, 42, 174–75,
        185*n*
solar wind, 57–58, 147, 165, 167, 196
    magnetosphere and bowshock,
        91–92, 170–71
Soter, Steven, 44
Soviet Union, 9, 46
    Halley's Comet, 221
    Mars exploration, 12
    radio frequencies, 176–77
    Sputniks, 11
    Venus exploration, 12–13
sun (*see also* Solar listings), 16, 165
Suomi, Vern, 233–34

tectonics, 110, 156
Tempel, Wilhelm, 18
Terrile, Richard, 198, 199, 213, 218,
    228, 229, 238
Tethys, 196, 201, 227, 247
    composition, 210
    cratering, 207–8, 242
    evolution, 208, 258
    and ring formation, 36–37
    terrain, 202, 207–8, 242, 258
Titan, 42–43, 129, 188, 211, 255
    approaches, closest, 202–4, 239
    atmosphere, 53, 55, 168, 175, 176,
        194–95, 202–4, 208–10, 212,
        239, 255

biological activity, potential for, 43,
    195, 204, 209, 255–57
blandness, 203
composition, 208–9
evolution, 210–11, 232
orbit, 66, 176
Pioneer mission, 168, 174 ff.
radio emissions, 210, 238–39
Saturn, interactions with, 168, 175,
    209, 212–13, 239
size, 210
temperatures, 176–78, 195, 204,
    208, 210–11
torus, 194
trajectories to, 65–66, 67, 84, 176,
    194, 202
volcanism, 43
Titan Centaur booster, 64
Tomasko, Martin, 176
Trainor, James H., 178–79
trajectories, 63–67, 84, 163–65, 172,
    202, 230, 233, 245
    corrections and adjustments,
        in-flight, 84–85, 133, 141
    gravity assists, 63, 115, 163, 172
Tyler, Len, 216, 237, 255

ultraviolet spectroscopy, 53–54, 92–93
Uranus, 39, 40
    hydrazine fuel for flyby, 84–85, 133
    observations from Earth of, 11, 113
    rings, 113, 114, 174
    trajectories to, 63–64, 65, 67, 164,
        172, 229
    travel time to, 63
    Voyager missions, 14, 48, 133, 182,
        222, 246–47, 260
Urey, Harold, 31
U.S.S.R., *see* Soviet Union

vacuum, Jupiter's magnetosphere, 147
Van Allen belts, 27, 165, 166
Van Allen, James, 27, 166, 167, 170,
    173
Van der Woude, Jurrie, 103, 122, 142
Venera spacecraft, 12–13
Venus, 39, 185
    atmosphere, 10, 129

271

INDEX

Venus (*cont.*)
  landings, unmanned on, 12–13, 14
  magnetic field, 12
  Mariner mission, 12
  observation from Earth of, 10
  Orbiting Image Radar mission, 182, 221, 259
  Pioneer mission, 14, 167, 177
  Soviet exploration of, 12–13
  temperature, 12
Viking missions and spacecraft, 62
  Mars, 13–14, 32, 50–51, 191
  moon landings, 13–14, 47
  press coverage, 126
Vogt, Rochus E., 59
volcanism
  Earth, 116, 158, 159
  Io, 106, 115–16, 134, 137, 142, 145, 158–61
  Titan, 43
Voyager missions and spacecraft (*see also* names, subjects), 3–5, 14, 48 ff., 140, 141, 170, 259 ff.
  construction and electronics of spacecraft (*see also* Instrumentation; subjects), 60–63
  costs and funding, 48–49, 222–24
  energy sources, 61–62
  evaluation and interpretation of findings, 114, 117, 129, 144–61, 166, 218–19, 254–58
  fate of spacecraft, 11, 67–68, 81–82
  Galileo mission and, 150–51, 182
  goals, scientific, and instrumentation for, 49–68, 114, 165, 224–26
  hydrazine fuel, 84–85, 133, 247
  launches, 64, 66, 83–84
  media and, *see* Media, coverage by, 90
  message records carried by, 70–82, 114
  optical navigation system, 66–67
  problems and failures, 84–86, 107, 108, 131–32, 133, 138, 141, 224–25, 246–50, 258
  staff (*see also* names), 50 ff.
  symposium, 102

Waldheim, Kurt, 77
Waldrop, Mitch, 103, 121
Warwick, James W., 54–55, 218
Wildt, Rupert, 23, 25
Wilford, John Noble, 121
Williams, Gareth, 148–49, 193–94
Wolfe, John, 166, 167, 170, 171, 178, 179

Young, Tom, 15, 164